£7-50
1500511686
[SF 396.5 KID)
AF844827

DIGESTION in the PIG

D.E. KIDDER, B.Sc., Ph.D., M.R.I.C., M.I.Biol.
Lecturer, Department of Veterinary Medicine, University of Bristol

and

M.J. MANNERS, N.D.A., M.Sc.(Agric.), Ph.D.
Lecturer, Department of Animal Husbandry, University of Bristol

SCIENTECHNICA BRISTOL

COPYRIGHT NOTICE

© *D.E. Kidder & M.J. Manners 1978*

All Rights Reserved. No part of this publication may be reproduced, stored in a retrieval system, or transmitted, in any form or by any means, electronic, mechanical, photocopying, recording, or otherwise, without the prior permission of the copyright owners.

CIP Data
Kidder, Douglas Edwin
Digestion in the Pig.
1. Swine — Physiology 2. Digestion
I. Title II. Manners, Michael Jackson
636.4'08'923 SF396.5

ISBN 0 85608 022 5

Printed by Kingston Press, Melcombe Road,
Oldfield Park, Bath BA2 3LR. Tel. 21948.

We feed our horses hay and oats,
With grass for cows and sheep and goats.
Chickens look for grain to eat,
While ducks find worms, and dogs get meat.
Cats have meat and milk and fish;
To each, its own peculiar dish.
Some are fussy, others not,
But pigs, of course, will eat the lot.

Acknowledgements

The collection of information for this book would not have been possible without the existence of the Agricultural Research Council Index of Current Research on Pigs, collated by Dr. R. Braude at the National Institute for Research in Dairying (N.I.R.D.) and the Intestinal Abstracts produced by the University of Sheffield Biomedical Information Service.

We are grateful to Dr. A. G. Low of the N.I.R.D. and Dr. R. F. Harvey of the Department of Medicine, the University of Bristol for friendly advice and assistance, and to the authors whose figures and tables we have been permitted to reproduce as acknowledged on the appropriate pages, and the respective journals and publishers, acknowledged in the corresponding references. We are also indebted to the University of Bristol Publications Committee for financial support.

We should like, also, to thank Mr. K. Howard for preparing many of the figures, Messrs M. Parsons and J. Conibear for photography and Miss A. Norris for typing and retyping the manuscript.

We wish to apologise for any errors and misrepresentations, but are confident that these will be brought speedily to our attention so that they can be corrected in any subsequent edition.

UNIVERSITY OF BRISTOL
VET. SCIENCE

PREFACE

The digestive system of the pig is noteworthy in that it contains almost all the features found in the digestive system of any mammal. Even carbohydrate fermentation in an oesophageal portion of the stomach, a feature fully developed in the ruminants, is present in a limited form in the pig. A study of digestion in the pig is thus a sound basis for its understanding in any other mammal. It is this comprehensive and flexible digestive system which gives the pig its capacity to cope with an extremely wide range of feeding stuffs and has thus formed the traditional basis of the usefulness of the pig to man.

Digestion in the pig has been studied by agricultural scientists interested in food utilization, by physiologists who have found the pig to be a useful experimental model for particular studies, and by biochemists and pharmacologists who have found pig slaughterhouse material to be a valuable source of digestive enzymes and gastrointestinal hormones. The results obtained have consequently been scattered among journals in very different fields, and written in a variety of languages. The present book is an attempt to bring this work together. It gives detail on the different digestive processes and shows how the various facets contribute to the functioning of the system as a whole. By showing the present state of knowledge, it should help research workers to concentrate on those aspects where little or no work has been done, and where our admitted ignorance, or our misconceptions based on a few doubtful studies, hinder correct interpretation of work on related topics.

From such a collection of information, two points emerge very clearly. One is the progressive changes with age of almost every aspect of the digestive system, and the other is the relationship between the findings of the physiologist on organ function and those of the nutritionist on growth performance. We have tried to present our material in a way which brings out these points.

While the book is designed primarily for research workers studying the digestive physiology of the pig, it will be of interest to those studying nutrition of the pig or the digestive systems of other species.

CONTENTS

CHAPTER 1
DIMENSIONS OF THE TRACT

INTRODUCTION

The structural anatomy of the digestive system of the pig is described and illustrated in textbooks of veterinary anatomy (e.g. Ellenberger & Baum, 1943; Nickel, Schummer, Seiferle, 1973; Sisson & Grossman, 1975) and will not be dealt with in this book. The organs are illustrated diagrammatically in Fig. 1.1. In the live animal, the stomach, the duodenum and the large intestines have fairly constant positions. The small intestine from the attachment of the pancreas, to the terminal ileum at the junction of the caecum and colon, is very mobile, being only held by its very flexible mesentery. While the small intestine will occupy a specific region of the abdominal cavity, any individual part of it may move quite considerably.

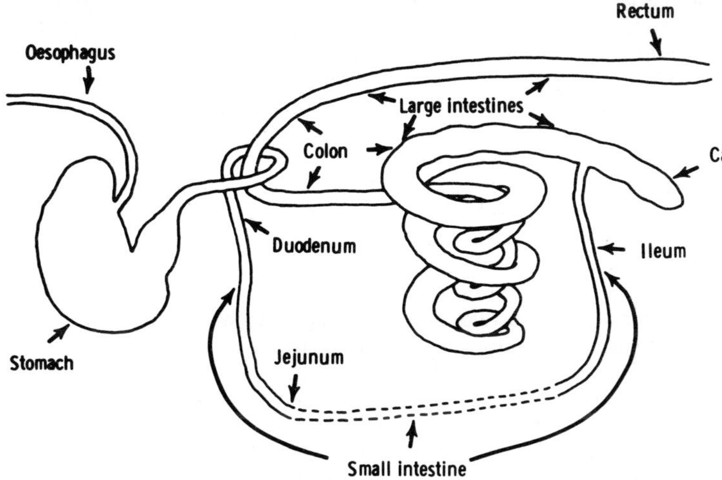

Fig. 1.1 Diagrammatic representation of the alimentary tract of the pig

Work on the dimensions of the tract is less adequately covered in anatomy textbooks and will be reported in more detail here.

There cannot be a precise value for the capacity or length of any part of the alimentary tract, as in life the tract is almost continually in motion, with the dimensions of each part altering considerably, sometimes from one second to the next. Almost all the measurements of the dimensions of the alimentary tract have necessarily been made post-mortem and will not correspond with those in life. It was shown by Hirsch, Ahrens & Blankenhorn (1956) in man, and by Leigh-Browne & Harpur (1975) in the pig, that the post-mortem length of the intestine, as measured, is appreciably greater than that during life. Laplace (1970) showed that the method of slaughter affected the post-mortem measurement of intestinal length. Measurements of capacity are also materially affected by the degree of muscle relaxation, so post-mortem values cannot be regarded as a measure of the capacity in the live animal. Nevertheless, post-mortem measurements can be of value in comparing the dimensions of the organs of different animals, provided they are made under comparable conditions.

The limitations of post-mortem measurements do not, of course, apply to weight, but, in all the dimensions (capacity, length and weight), the results of different teams of workers have been remarkably similar. A feature of all of them is a relatively rapid growth between five and eight weeks of age. This is more marked in the large intestines than in the other sections of the gut, and more in the weight than in the other dimensions.

CAPACITY

The most satisfactory way of measuring the capacity of any section of the gastro-intestinal tract is to immerse the organ in water or saline with one end closed, and then to find the amount of water or saline required to fill the organ. Kvasnitskii (1951) showed that, with this method, the hydrostatic pressure used had a marked effect on the results, particularly with the stomach, and gave results for the capacity of the stomach of six pigs measured at hydrostatic pressure differences between the inside and outside of 10 and 40 cm of water (Table 1.1). He also showed that there was variation in apparent stomach capacity obtained by slaughtering at different times after feeding (Table 1.2).

Using a standard hydrostatic pressure of 10 cm of water and a standardised feeding regime, Kvasnitskii measured the post-mortem capacity of the stomach, small and large intestines of pigs of ages from 1 day to 4 years (Table 1.3). Using the same hydrostatic technique, Jezková (1962) made a study on sow-reared piglets, covering the age-range of 1–26 days in more detail (Table 1.4). Vodovar, Flanzy & Francois (1964) compared the hydrostatic method of measuring capacity of the small intestine with results calculated from the length and diameter, and found the two methods to agree to within 8%. Using length and diameter measurements on pigs from birth to 3 years of age, they calculated the small intestine capacities (Table 1.5).

If we compare the mean values for capacity of the various organs obtained by Kvasnitskii, by Jezková and by Vodovar, Flanzy & Francois (Figs. 1.2, 1.3 and 1.4) we find that in spite of the factors which can cause considerable variation in this type of measurement, the various authors obtained quite similar results.

Table 1.1 The apparent capacity of the stomach of the pig when measured with different pressures of water (from Kvasnitskii, 1951)

Weight of animal (kg)	Capacity of stomach (litres) under pressure of water column	
	10 cm	40 cm
97	3.8	8.0
100	4.6	7.2
99	4.0	7.4
158	4.6	9.0
154	5.8	11.8
154	7.0	12.4

Table 1.2 *Effect of time between feeding and slaughter on capacity of the digestive organs (mean values on 10 pigs in each group) (from Kvasnitskii, 1951)*

		Capacity (litres)		
Wt. (kg)	Interval between feeding and slaughter (hours)	Stomach	Small intestine	Large intestine
155	5	6.0	19.4	18.8
154.6	15	3.7	17.7	18.6

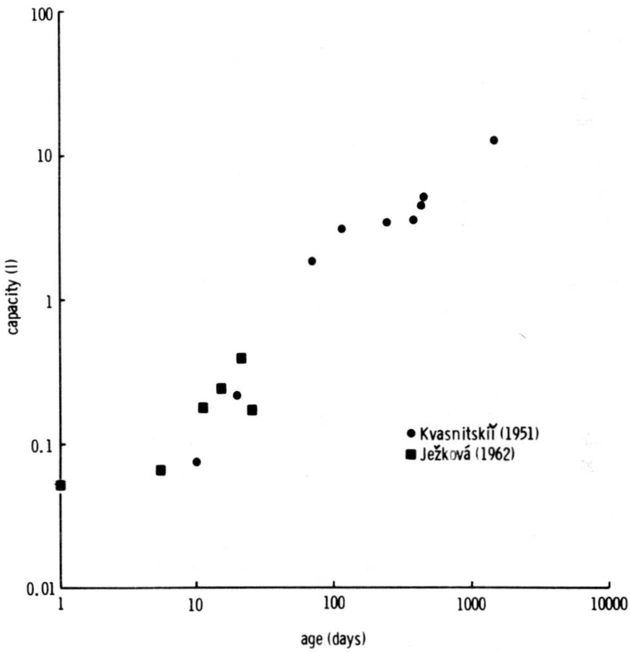

Fig. 1.2 *The mean capacities of stomachs of pigs of different ages as found by different workers. Both age and capacity are on logarithmic scales*

Table 1.3 *Mean weights and capacities of the stomach and weights, capacities and lengths of the intestines of pigs of different age groups (from Kvasnitskii, 1951)*

		Stomach		Small Intestine			Large Intestine		
Mean weight (kg)	Age (days)	Weight (g)	Vol. (litres)	Weight (g)	Vol. (litres)	Length (m)	Weight (g)	Vol. (litres)	Length (m)
1	1	4.5	0.025	40	0.1	3.8	10	0.04	0.8
2	10	15	0.073	95	0.2	5.6	22	0.09	1.2
3	20	24	0.213	115	0.7	7.3	36	0.1	1.2
18	70	232	1.815	996	6.0	16.5	458	2.1	3.1
32	115	360	2.50	1180	10.7	18.0	714	6.6	4.3
69	208	685	3.17	1670	13.3	18.8	1380	11.7	5.4
103	255	754	3.40	1530	14.1	18.8	1280	10.1	5.0
152	380	980	3.55	2510	20.6	23.7	2010	15.7	6.8
156	428	844	4.56	2323	19.1	22.4	2184	18.1	6.1
154	449	980	5.16	2310	17.4	21.2	1970	17.2	6.3
270	4 years	1430	12.68	1998	22.6	22.9	2790	25.6	7.5

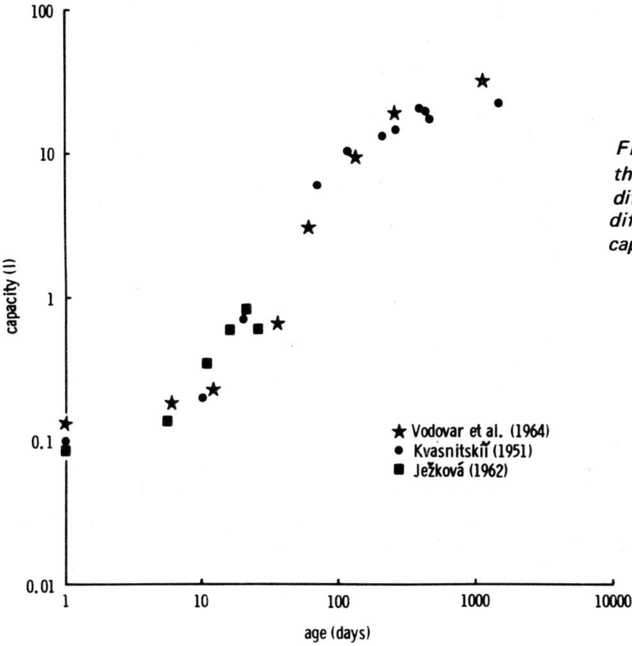

Fig. 1.3 Mean capacities of the small intestines of pigs of different ages as found by different workers. Both age and capacity are on logarithmic scales

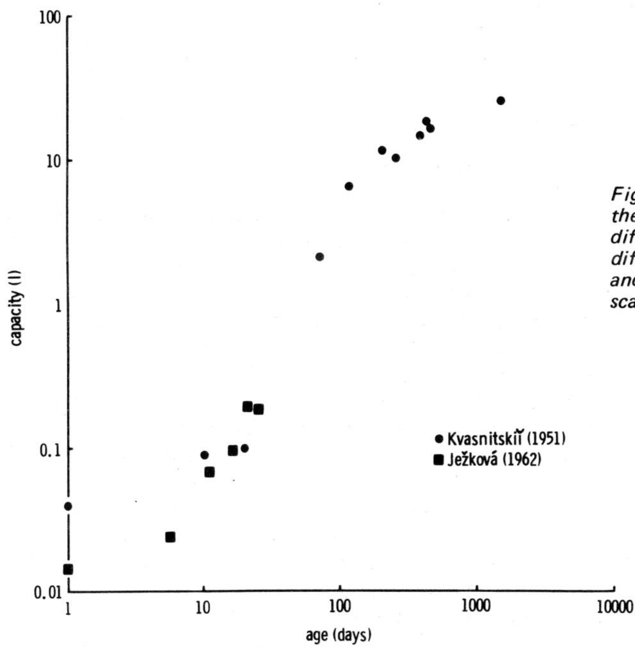

Fig. 1.4 Mean capacities of the large intestines of pigs of different ages as found by different workers. Both age and capacity are on logarithmic scales

Table 1.4 *Capacities of the stomach and intestines of sow-reared piglets up to 26 days of age (from Jezková, 1962)*

Age (days)	No. of animals	Stomach	Volume (cm^2) Small Intestine	Large Intestine
1	5	52 ± 14	88 ± 14	14 ± 6.0
5–6	5	65 ± 10	138.6 ± 35	23 ± 1.1
11	6	179 ± 33	350 ± 25	66 ± 10
16	5	240 ± 34	587 ± 42	94 ± 14
21	4	390 ± 128	824 ± 142	190 ± 107
25–26	5	172 ± 95	585 ± 324	184 ± 117

Effect of Diet on Capacity

There have been a number of reports demonstrating that the bulk of the food material ingested affects the subsequent capacity of the gastrointestinal tract.

Kvasnitskii (1951) compared pigs reared from 3 months of age (18 kg live weight) on either bulky or concentrated diets, and at slaughter, found the mean total capacities of the digestive organs to be greater on the bulky diets. Starovoitov (1956) compared the capacities of the digestive organs at slaughter of 4 groups of pigs which had had different supplements during the suckling period and found that the capacities of the stomach, small and large intestines were greater where more supplement was given, and still greater when this contained bulky foods.

Shearer & Dunkin (1968) found that pigs fed to 120 lb weight on high-lactose diets had a much larger caecum than those on a low-lactose diet or one without lactose. As described in Chapter 6, lactose, when fed at high levels, is incompletely absorbed in the small intestine of adult pigs, so that there is osmotic retention of water and a consequent increase in the volume of the contents entering the caecum.

LENGTH

Leigh-Browne & Harpur (1975), using a plastic tube inserted through a cannula to measure the in vivo length on a conscious pig, found this to be about half the post-mortem length. This is about the same ratio of in vivo to post-mortem length as was found in man by Hirsch, Ahrens & Blankenhorn (1956). While Nickel (1933) found that a wide tube passed in vivo caused rucking up and shortening of the intestine, Hirsch et al. and Leigh-Browne & Harpur established that this was not the case in their studies. Laplace (1970), comparing lengths of intestinal loops in fasted, anaesthetized pigs with measurement made post-mortem on the same animals, found the in vivo measurements to be longer, but this might be related to fasting and anaesthesia. He also found that the method of slaughter made a difference to the post-mortem length. The shrinkage after removal of the intestine from the pig can be readily observed especially in empty sections of the small intestine. Results of measurements must therefore be regarded only as approximate and as comparisons rather than absolute values.

Figures for the length of the small and large intestines from birth to 4 years of age were given by Kvasnitskii (1951) (Table 1.3) and by McCance (1974) and

Table 1.5 Lengths and capacities of the small intestines of pigs (from Vodovar, Flanzy & Francois, 1964)

	Pigs		Small Intestine			
				Diameter		
Number	Age (days)	Weight (kg)	Length (m)	Duodenal end (cm)	Ileal end (cm)	Capacities (ml)
22	birth	0.950–1.150	2.15 (2.07–2.24)	0.65 (0.60–0.68)	0.65 (0.59–0.68)	72
12	1	1.100–1.250	3.05 (2.94–3.14)	0.72 (0.67–0.76)	0.72 (0.65–0.75)	130
12	6	1.840–2.300	3.85 (3.69–3.97	0.78 (0.73–0.80)	0.78 (0.71–0.80)	183
8	12	2.500–2.800	4.50 (4.33–2.62)	0.8 (0.77–0.85)	0.8 (0.75–0.83)	226
5	30–40	7–9	6.55 (6.31–6.73)	1.62 (1.49–1.72)	1.58 (1.48–1.69)	662
3	55–65	15–17	9 (8.70–9.35)	2.4 (2.1–2.7)	2.2 (1.9–2.5)	3110
15	120–140	45–55	13 (12.46–13.46)	3.2 (2.75–3.55)	3 (2.7–3.4)	9744
35	200–230	95–105	16.50 (15.87–17.27)	4.2 (3.6–4.7)	3.6 (3.3–4.1)	19182
2	3 years	223 & 245	18.40 & 19.60	4.9 & 5.1	4.6 & 4.8	32945

for the small intestine by Vodovar, Flanzy & Francois (1964) (Table 1.5). More detailed figures over the age range of 1–26 days were given Jezková (1962) (Table 1.4). Values for newborn, 1 and 10 day old pigs have been given by Widdowson, Colombo & Artavanis (1976).

Fig. 1.5 shows the results of Kvasnitskii, Vodovar et al., Jezková, McCance, and Widdowson et al., for the small intestine, plotted with a logarithmic age scale. From 20 days onwards, the measurements given by Vodovar et al. are rather lower than those of the other three authors, which show remarkable agreement. Fig. 1.6 shows the results of Kvasnitskii, Jezková, McCance and Widdowson et al. for the large intestine. Again, in spite of all the possible causes of variation, the measurements are in good agreement.

WEIGHT

The total tissue weights of various sections of the digestive tract in pigs of different ages have been reported by McMeekan (1940) and Kvasnitskii (1951) and for newborn, 1 day and 10 day old piglets by Widdowson, Colombo and Artavanis (1976). Figs. 1.7, 1.8 and 1.9 illustrate diagrammatically the results of these workers. The similarity of the results of the various workers is evident, as in the results on length and volume. The relatively rapid increase in weight of the large intestine between 5 and 8 weeks of age is noteworthy.

A significant difference has been demonstrated by Miniats & Valli (1973) between the weights of the intestines of germ-free and conventional young pigs, with specific-pathogen-free pigs being intermediate. The intestines of the germ-free pigs had thinner walls and were consequently lighter.

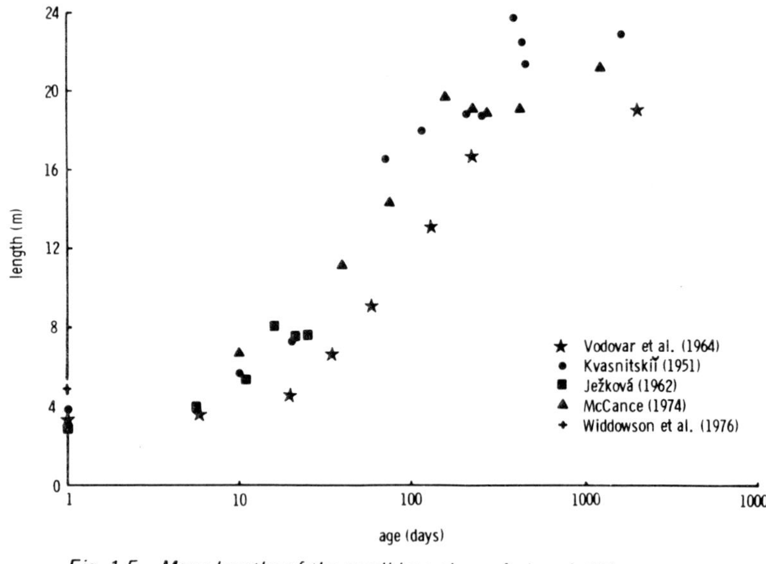

Fig. 1.5 Mean lengths of the small intestines of pigs of different ages as found by different workers. Age is plotted on a logarithmic and length on a linear scale

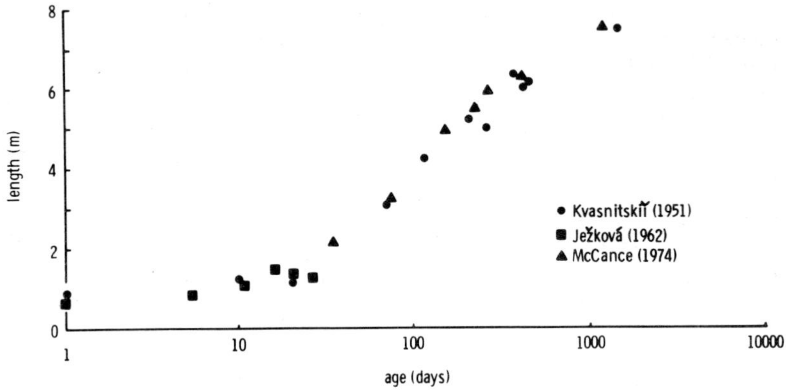

Fig. 1.6 Mean lengths of the large intestines of pigs of different ages as found by different workers. Age is plotted on a logarithmic and length on a linear scale

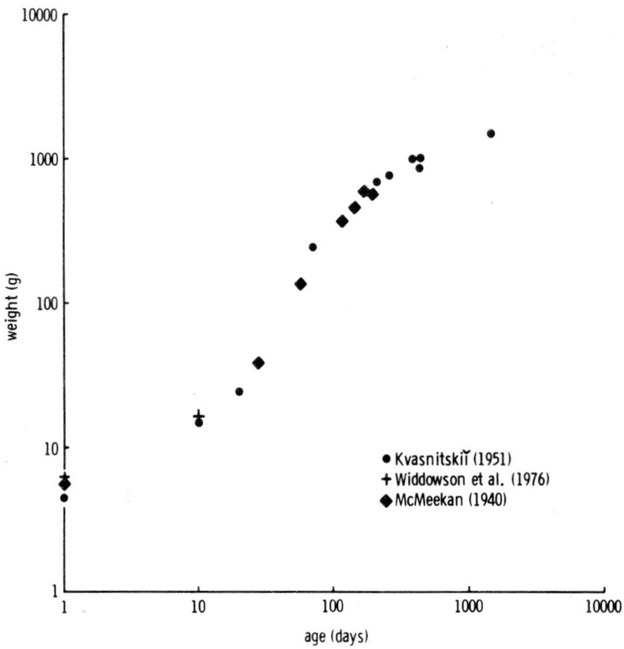

Fig. 1.7 Mean tissue weights of stomachs of pigs of different ages as found by different workers. Both age and weight are plotted on logarithmic scales

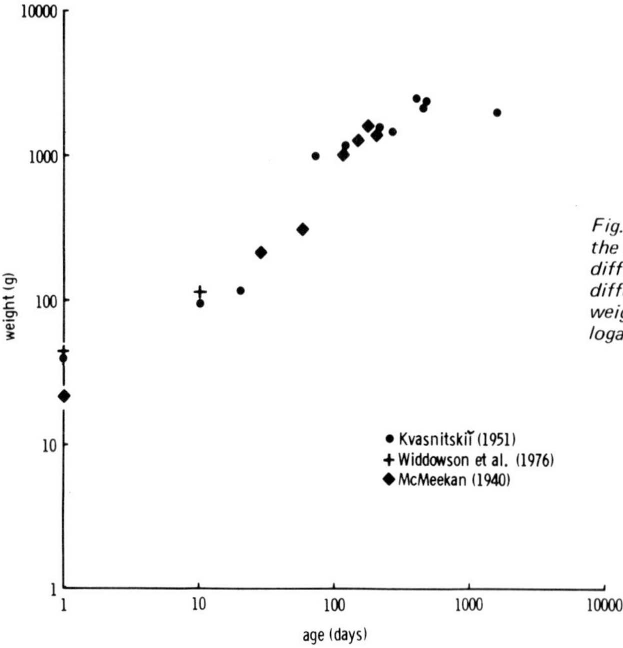

Fig. 1.8. Mean tissue weights of the small intestines of pigs of different ages as found by different workers. Both age and weight are plotted on logarithmic scales.

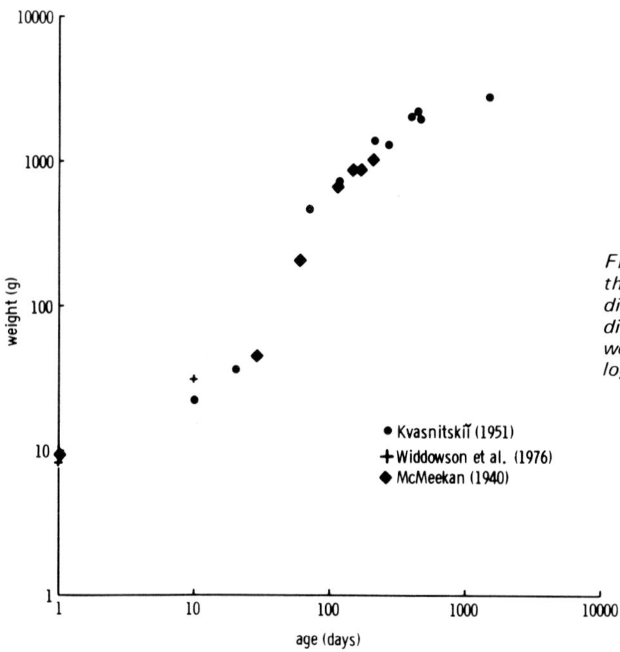

Fig. 1.9 Mean tissue weights of the large intestines of pigs of different ages as found by different workers. Both age and weight are plotted on logarithmic scales.

REFERENCES

Ellenberger & Baum (1943) Handbuch der vergleichenden Anatomie der Haustiere Edn 18. Zietzschmann, O., Ackerknecht, E. & Grau, H. pp. 451–456. Springer–Verlag, Berlin.

Hirsch, J., Ahrens, E.H. & Blankenhorn, D.H. (1956) Measurement of human intestinal length in vivo and some causes of variation. *Gastroenterology, 31,* 274–284.

Jezková, D. 1962) Délkové, váhové a objemové zmeny strev a objemové zmeny zaludku u selat v raném postnatalnim údobi. *Vet. Cas., 11,* 499–508.

Kvasnitskii, A.V. (1951) Voprosy fiziologii pishchevareniya u svinei. Sel'khozgiz, Moscow.

Laplace, J.P. (1970) Mensuration in vivo et postmortem, et croissance de l'intestin grêle chez le porc. *Annls Zootech., 19,* 465–469.

Leigh-Browne, G. & Harpur, R.P. (1975) Intestinal intubation in the unanaesthetised miniature pig: technique for sampling and measurement of gut length per fistulam. *Res. vet. Sci., 18,* 6–14.

McCance, R.A. (1974) The effect of age on the weights and lengths of pig's intestines. *J. Anat., 117,* 475–479.

McMeekan, C.P. (1940) Growth and development in the pig with special reference to carcass quality characters. *J. Agric. Sci., 30,* 276–343.

Miniats, O.P. & Valli, V.E.O. (1973) The gastrointestinal tract of gnotobiotic pigs. *in* Germfree research. Biological effect of gnotobiotic environments. Ed. Heneghan, J.B. pp. 575–583. Academic Press, New York.

Nickel, R. (1933) Über die Ermittlung der Länge und Lage des Verdauungskanals. Untersuchungen an toten und lebenden Hunden zur Gegenüberstellung der intravitalen und postmortalen Verhältnisse. *Z. ges. exp. Med., 91,* 193–243.

Nickel, R., Schummer, A. & Seiferle, E. (1973) The viscera of the domestic mammals. pp. 137–146. Paul Parey, Berlin & Hamburg.

Shearer, I.J. & Dunkin, A.C. (1968) Caecum enlargement in the lactose-fed pig. *N.Z. J. agric. Res., 11,* 923–926.

Sisson, S. & Grossman, J.D. (1975) The anatomy of the domestic animals. Edn 5. Ed. Getty, R. vol. 2, pp. 1268–1282. W.B. Saunders Co., Philadelphia.

Starovoitov, A.M. (1956) Effektivnost' podkormki porosyat-sosunov zernovymi i sochnymi kormami. *Zhivotnovodstvo., 6,* 46–51.

Vodovar, N., Flanzy, J. & Francois, A.C. (1964)Intestin grêle du porc. I. Dimensions en fonction de l'age et du poids, étude de la junction du canal choledoque et du canal pancréatique a celui-ci. *Annls Biol. anim. Bioch. Biophys., 4,* 27–34.

Widdowson, E.M., Colombo, V.E. & Artavanis, C.A. (1976) Changes in the organs of pigs in response to feeding for the first 24 hours after birth. II. The digestive tract. *Biol. Neonate, 28,* 272–281.

CHAPTER 2
PASSAGE OF FOOD

Published work on the passage of food through the alimentary tract of the pig has been reviewed by Laplace (1975) in a comparative study covering various mammals, including the pig. Work in this field can be divided into studies of the nature of the movements, and studies of the rate of passage of the contents of the lumen.

MOVEMENTS

Studies of the movements of the gastrointestinal tract of the pig have been made using radiography, electromyography either with direct electrical connections or with radio-telemetry, and by insertion of balloons to measure pressure changes.

The Stomach and Duodenum

The best general picture of the movements is obtained by radiography, monitoring with a fluorescent screen. Detailed descriptions of the stomach movements of 3–5 month old pigs observed in this way were given by Neimeier (1940, 1941).

Neimeier observed three layers in the stomach; the bottom layer containing the solid matter occupied one half to two thirds of the stomach, and above this there was a fluid layer and a layer of gas. When food was given, peristaltic waves set in while the pig was still eating. The contractions appeared to come from the base and move upwards towards the pylorus but did not always cause food to pass into the duodenum, the opening of the pylorus occurring irregularly. Vigorous stomach activity started at the beginning of food intake and usually continued for three quarters of one hour, followed by a rest for half an hour broken only by occasional waves, then activity restarted, but less vigorously. After 3 hours or less, the stomach usually rested altogether with just an occasional peristaltic wave. The stomach did not empty completely and a considerable amount of food residue remained even after a 24 hour fast.

In the duodenum the activity was related to the gastric peristalsis. For the first three quarters, to one hour after a meal, peristalsis was very active, especially in the first half hour when 5–8 food portions per minute passed into the duodenum. Then there was a reduction in activity for half an hour before vigorous peristalsis set in again, with 3–4 food portions passing through the duodenum every two minutes. The portions had a cigar shape and took 4–6 seconds to pass through the duodenum which appeared to be closed except where one of these food portions was passing through.

Observations using electromyography, and pressure studies using balloons have, in general, supported and amplified these observations in relation to the type of contractions and the rest periods.

Laplace (1972a), using electromyography, found a region in the cardia with little rhythmic activity. In the antrum in the fasting pig, there were rhythmic contractions moving with increasing amplitude towards the pylorus, the frequency varying from one individual to another within the range of 3.4 to

4.5 contractions per minute. This type of rhythm was also found in the duodenum at the level of the bile duct insertion (Laplace, 1972b). Laplace (1974a) found that the only effect of feeding on this rhythm was a slight speeding up of the frequency during the meal.

Hahn (1975), who used a miniature telemeter implanted in the stomach, and so was able to monitor the movements without disturbing the animal, classified the movements into three types:—

1) A "fast" rhythm due to peristaltic contractions of the stomach at 3 to 5 contractions per minute.
2) A "slow" rhythm in which one or more contractions from the fast rhythm were missing.
3) Breaks in contractile activity of 2—60 minutes — the "rest" periods.

The fast rhythm occurred immediately after feeding, was more rapid and had a higher amplitude after a large feed, and returned to its usual rate as the stomach emptied. The slow rhythm which sometimes followed came about six hours after a large feed. The rest periods started about 9 hours after a large feed and continued at one to one and a half hour intervals, increasing in duration with time after feeding. About 15 hours after a large feed, the contractions increased in intensity and these "fasting" contractions gradually acquired a greater amplitude than the initial contractions after feeding.

These observations confirm other reports on individual features of the movements of the stomach and duodenum. Ruckebusch & Bueno (1976) using electromyography, reported continuous activity in the antrum and duodenum for a long time after a large feed and for a shorter time after a small feed. Plonait (1974) using telemetry, reported the contractions to last 15 seconds with a periodicity of 10—20 seconds.

The rest periods were observed by Kratinov & Kratinova (1932) using balloons to measure pressure in the lumen. Plonait (1974) reported the rest periods to be longer and more frequent at night.

The Intestines

The small intestine

The nature of the movements in the small intestines were best described by Neimeier (1940, 1941). He observed that the portions of food slowed down on reaching the jejunum and could no longer be followed individually. Initially there were no large movements, then one and a quarter to two hours after feeding, vigorous activity set in with two main types of movement.

1) A finger-thick loop would rise out of the mass of loops and fall back over and over again for a few minutes or as much as half an hour, the contents of the loop appearing to move to and fro. Then it would settle down or the contents be moved on by a peristaltic wave. Then next to it one or more other loops would be thrown about in the same way.

2) At the same time large peristaltic waves were passing. The contents of a loop would suddenly move upwards to the ventral part of the loop and pass on, or sometimes stay at the top of the loop for up to 10 minutes and then pass on. Sometimes there were 5—6 peristaltic waves per minute, and sometimes long periods with only an occasional one.

Sometimes gentle rocking of the whole jejunum appeared for a short time, and in the dorsal loops some creeping and pendulum movements. Rhythmical segmentation also occurred in which a succession of deep constrictions, lasting only a short time, gave the momentary appearance of a string of pearls.

After 3–4 hours the up and down movement of the jejunal loops gradually stopped from the cranial end caudally, and the loops emptied again from the cranial end. After 6–10 hours the jejunum had usually emptied.

The ileum stood out from the jejunal convolutions as a cord in the region of the last bow ribs, casting a continuous shadow. The contents ran up, and sometimes remained half way to ostium ileocaecocolicum for a few seconds to a minute before falling back or passing on. Sometimes they went straight on and passed into the large intestine with or without a pause at the ileocaecal valve. The speed varied greatly; the wave might travel 1–2 cm in a second or traverse the entire section in seconds. Transfer of contents occurred in pulses with intervals which varied from a few seconds to 4–8 minutes. About 5 hours after feeding, activity of the ileum subsided.

Studying movements in the small intestine by electromyography, Laplace (1972a) recorded three types of activity; 1) local irregular activity, 2) segmented activity affecting a large or small region, with activity passing distally along the segment, and 3) rapid peristaltic waves.

The irregular activity was commonest in the young pigs (30 kg) and the segmental and rapid peristaltic waves commonest in the older pigs (70–90 kg). This difference was more marked if the younger pigs were fed a growing ration and the older pigs a finishing ration. A basal rhythm was also present, which had a mean frequency at the duodenum in the older pigs of 17.06 waves/minute, and in the younger pigs 18.33 waves/minute. This frequency diminished progressively down the small intestine to 11.2 waves/minute by the distal ileum.

Ruckebusch & Bueno (1976) considered that in the jejunum of the pig after large feeds, the general pattern of activity resembled that in carnivores, with the regular pattern of rhythmical contractions replaced by more continuous activity. With small feeds, or ad lib. feeding, the pattern more closely resembled that in ruminants, with the fasting rhythm not greatly altered by the feeding.

The large intestine

Trautman & Ascher (1940) made a detailed study of the movements in the large intestine of pigs by infusing contrast medium into the lumen of the large intestine at various points via a tube through a caecal fistula. They divided the type of movements observed into mixing movements and transport movements and further subdivided the mixing movements seen into inversion movements and tonic contractions. The inversion movements consisted of rhythmic contractions at one place, where, unlike the contractions in the small intestine, they restarted again in the same place after stopping. The tonic contractions often lasted a minute or more and frequently occurred along the whole length of the colon dividing it into segments.

The transport movements were divided by Trautman & Ascher into weak and strong peristaltic movements, and large colon movements where a length of 15–30 cm contracted and went into active movement without ring contractions being observed. Both long and circular muscle appeared to be involved in the large colon movements which occurred in all parts of the colon but especially near the rectum.

The movements were also described by Neimeier (1940, 1941), the following being a summary of his account.

The contents of the ileum nearly always passed first into the caecum. Occasionally they passed first into the colon, and then into the caecum.

The most frequent movement observed was rhythmical segmentation. Usually there were several shallow (1–2 cm deep) constrictions starting at the caudal edge giving a toothed appearance to the caecum and adjacent part of the colon. Deeper constrictions in which the contents were separated, were less common in the caecum. Tonic contractions occurred, sometimes lasting sufficiently long for the caecum to take on a dumb-bell shape. There were weak and strong peristaltic waves, the strong ones usually starting in the upper third of the caecum and passing into the upper colon. Large contractions sometimes occurred with the caecum contracting to one third of its original length, often simultaneously with a peristaltic wave, so that the contents of the caecum emptied into the colon. In emptying, usually the upper third of the caecum widened, and a peristaltic contraction started distally and passed along to the colon moving the contents into the first coil where they fell ventrally. After 12–16 hours the caecum had usually shrunk to two thirds of its original length.

The shallow and the deep segmentations of the caecum usually continued into the upper colon, deep segmentation simultaneously at several places being typical of the upper colon. Strong peristaltic waves with complete separation of the contents occurred and were particularly common in the first loops of the colon, these having originated in the caecum. Weak peristaltic waves, where the lumen was merely narrowed, were less common. Tonic contractions also occurred, as in the caecum. The upper part of the caecum and colon always filled with gas. About 6–8 hours after a meal the activity of the large intestine subsided and there was a rest period only interrupted now and then by major movements.

Fourteen to sixteen hours after the meal, the first faecal pellets started to leave the colon. Completely formed balls with a definite spacing could be seen in the last terminal loops and passed slowly, 2–3 per minute cranially into the terminal bend. The movement was irregular with long pauses, and the distal end sections of the colon showed periodic pulsations. After the bend, the pellets took on a cigar shape. In front of the rectal entrance they closed up together to make a coherent faecal column which was united by successive ring contractions.

Gut Movements in the Piglet

Radiographic studies on 1–3 week old piglets by Kidder & Manners (1968, 1974) indicated that the gut movements in piglets closely resembled those reported in older pigs, the main difference being the much greater degree of emptying of the stomach of the piglet between feeds.

Starovoitov (1960), making pressure measurements with balloons, observed the same type of fasting contractions in the stomach of piglets as have been reported in older animals.

REGIONS OF DILATION AND CONSTRICTION

In his radiographic studies, Neimeier (1940) observed that the duodenal lumen appeared to be closed except for the few seconds at a time when a food portion was passing through. Kvasnitskii (1951), examining the intestines removed from pigs during slaughter studies, remarked on the way the small intestine of the pig consisted of lengths which were full of partly digested food alternating with lengths which were closed and almost empty. He found, in live pigs with intestinal fistulas, that the fistularised portion of the intestine would alternate between a state when it was filled with digesta and a state when it was empty.

Stevens & Kidder (1972) drew attention again to this phenomenon in the pig small intestine and, as described in Chapter 6, found differences in the mucosal enzymic levels between the full and empty sections of the small intestine.

Fossum & Liven (1974) examined the consistency and bacterial and enzyme content of such contents as could be found in the closed or constricted part of the small intestine, and showed that these contents were almost sterile, contained little or no pancreatic enzymes, and differed very much in consistency from the contents of the full sections proximal or distal to them. The contracted sections appeared to contain little other than intestinal mucus, while the full sections mainly contained partly digested food.

RATE OF PASSAGE

The rate of passage of food includes measurements of rate of travel, that is, the time taken to travel from one part of the digestive tract to another, and rate of flow, that is the volume passing a given point in unit time.

Rate of Travel

The overall time of passage from oral intake to anal excretion has been measured by putting markers in the food and recovering them from the faeces. The time to pass from one section of the gut to another has been studied with markers which are located either radiographically or by slaughter a given time after feeding, or on pigs fitted with re-entrant cannulas.

A problem with all these procedures is that different sized particles may travel at different speeds (Argenzio & Southworth, 1974) and the fluid in the digesta may travel much faster than the solid matter (Argenzio & Southworth, 1974; Clemens, Stevens & Southworth, 1975). Fig. 2.1 shows the results obtained by Argenzio & Southworth on the excretion of solid markers of three sizes, and a liquid marker. Nearly all the studies so far have been made using small solid particles as markers.

Overall time of passage

The markers used include barium sulphate detected radiographically, chromic oxide and ferric oxide, which have usually been detected chemically, and stained food particles and beads or pellets of various materials which were recovered mechanically from a slurry of the faeces, and counted.

Most workers reported the time after feeding when the marker first appeared in the faeces, and the time after which the faeces were free of marker. With growing or mature pigs most of the times reported were in the range 15–24

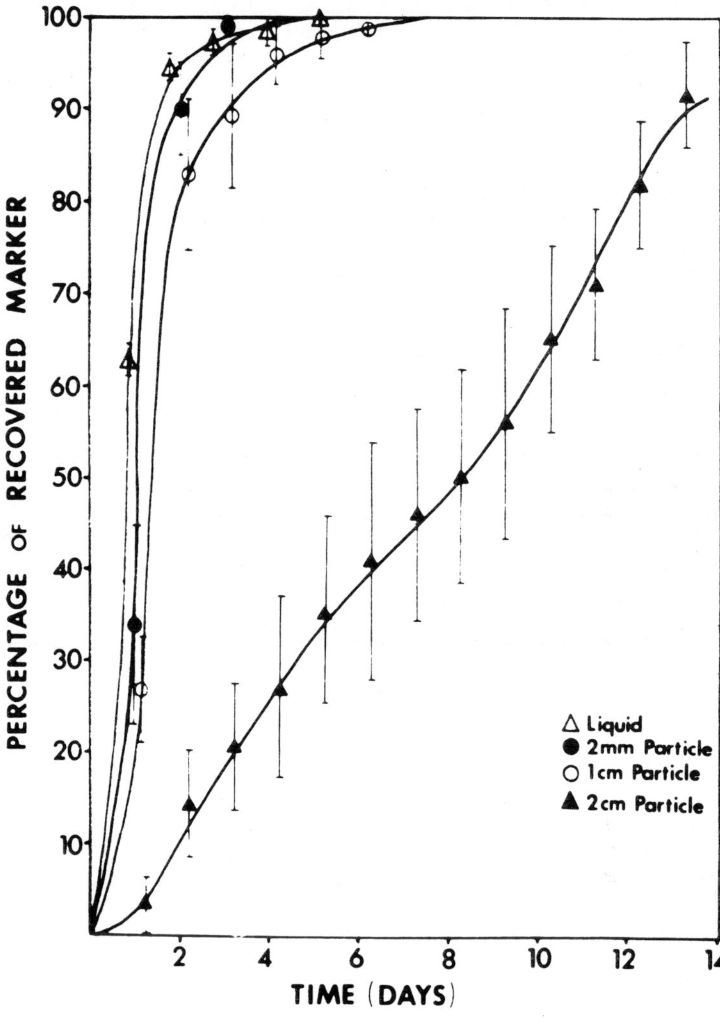

Fig. 2.1 Faecal excretion of liquid and particulate markers in 5 pigs fed control diet. Symbols represent percentage (±SE) of a given marker recovered in faeces. (Argenzio & Southworth, 1975)

hours from the initial appearance of the marker in the faeces (Falaschini & Carucci, 1934; Neimeier, 1940; Thurber, Hudman & Peo, 1965; Seerly, Miller & Hoefer, 1962; Green & Jewell, 1965). Slightly earlier initial times were reported by Castle & Castle (1957) who gave 10–24 hours and Krüger & Meyer (1958) who gave 10–17.5 hours.

With suckled piglets, Kidder & Manners (1968, 1974) found slightly longer times (24–30 hours). With early-weaned piglets, the time seems to be related to diet. Kidder, Manners & McCrea (1961), using a purified diet based on casein, and Maner, Pond, Loosli & Lowrey (1962) using a casein diet, obtained overall transit times of over 24 hours, like a suckling pig, while Maner, Pond, Loosli & Lowrey (1962) on a soybean protein diet, and Kidder & Manners (1974) on a commercial diet, obtained times of under 24 hours, resembling those in a mature pig.

The time taken, after ingestion of the marker, before the faeces are free of it, is much more variable. Times reported range from about 40 hours (Thurber, Hudman & Peo, 1965), through 50 hours (Green & Jewell, 1965), 54–80 hours (Krüger & Meyer, 1958) to the region of 78–90 hours reported by Falaschini & Carucci (1934), Castle & Castle (1956) and Horszczaruk (1962). The difference could be due both to the diet and to the sensitivity of the detection of the marker.

Kidder & Manners (1974) found overall excretion times of 2–4 days for suckled 1–3 week old piglets but shorter times for early weaned piglets on a commercial diet. Kidder, Manners & McCrea (1961) found times of 90–150 hours for early-weaned piglets on a purified diet.

The measurement of the initial appearance and, still more, the final disappearance of the marker is unsatisfactory, as we do not get a sudden disappearance of marker, but rather a gradual fall in its concentration. Because of this, Castle & Castle (1956, 1957) calculated the time taken for 5% and 95% of the ingested marker to be excreted by plotting a curve for recovery of marker from the faeces (Fig. 2.2). This technique has since been used by other workers and

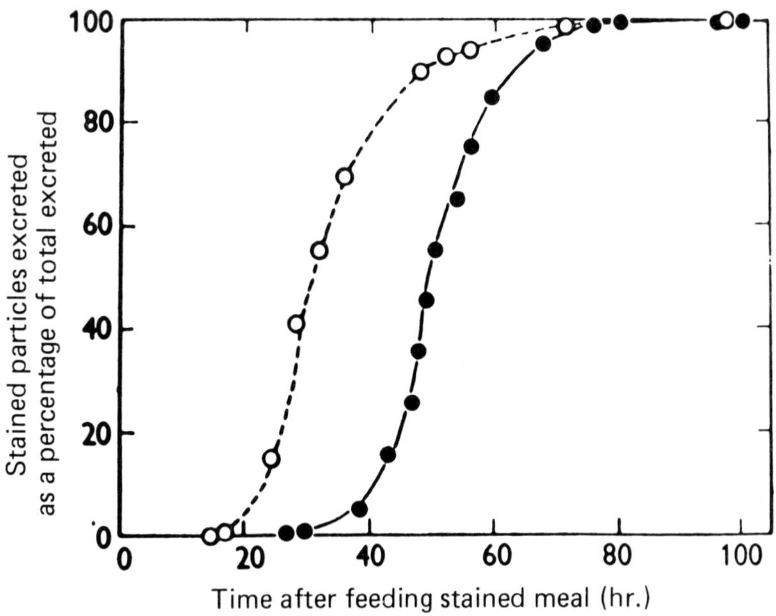

Fig. 2.2 Excretion curves of undigested residues from a feed of stained pig meal. ○—○, 94.5 lb. pig; •—•, 470 lb. sow (average of five experiments). (Castle and Castle, 1956)

Table 2.1 Time for 5% and 95% excretion of markers as reported by various workers

Authors	Marker	No. of individual passages on animals	Weight (kg)	Time (in hours) for: 5% excretion	95% excretion
Castle & Castle (1956)	Stained food particles	8 1	40 230	21 38	53 68
Castle & Castle (1957)	"	40	22–87	10.5–23.3	46.0–54.3
Seerly, Miller & Hoefer (1962)	"	108	80–90	17.6–23.9	51–81.4
Horszczaruk (1962)	Rubber pellets	9		18	48–60
Carillo & Benevides (1971)	Cr_2O_3	12	40	Molasses diet 18.8	66.7
				Maize diet 23.3	63.8
Marsboom, Temmerman & Symoens (1973)	Plastic pellets	32	24–30	16–24	88

Table 2.1 summarizes the results reported on this basis.

Several authors have studied the effect of changes of diet and feeding regime of the time of passage. Castle & Castle (1957) showed that the time of food passage varied inversely with the bulk of the food taken, whether this was effected by varying the water content or the dry-matter content of the mash. The same authors found, in 1956, that the morning feed remained in the tract for a shorter time than the afternoon feed, a finding supported by Krüger & Meyer (1958) but not by Seerly, Miller & Hoefer (1962).

Most workers (e.g. Castle & Castle, 1956; Krüger & Meyer, 1958; Seerly, Miller & Hoefer, 1962) have not found significant differences in time of passage between diets of different digestibilities.

Time of passage in various sections of the tract

Keys & DeBarthe (1974), using stained diets in fistularized pigs, measured the time taken between ingestion and maximum dye passage at the duodenal fistula, the ileal fistula and the anus with various cereal diets. Owing to the variability, they did not find significant differences between the diets. They found maximum dye passage times of 2.5–5 hours for the duodenal fistula, 6–9 hours for the ileal fistula and 38–45 hours for the faeces.

Various other workers have reported on the times of passage in individual organs as described below.

THE STOMACH

Emptying

The factors regulating the passage of food from the stomach into the intestines, "gastric emptying", has been studied mainly in the human and the dog, but also to a limited extent in other mammals including the pig. The same factors seem to control the process in the different species, and are described by Thomas (1957), Hunt & Knox (1968) and Cooke (1975).

The stimulus for emptying is the volume of material in the stomach, and initially the rate of emptying is related to this volume and consequently gives a volume/time emptying curve which is approximately exponential (Hunt & Macdonald, 1954; Hopkins, 1966). The continued emptying is then influenced by nervous and hormonal inhibitory effects.

Several of the gastrointestinal hormones described in Chapter 4, notably secretin, gastric-inhibitory peptide (GIP) and vasoactive intestinal polypeptide (VIP) will inhibit the gastric contractions, while motilin increases them, but it is not known yet which of them make significant contributions to the control of gastric emptying under physiological conditions. Cutting the nervous connections between the stomach and the duodenum reduces the inhibitory effects of food in the duodenum, so the nervous system also contributes to the control, but the extent of its contribution is not known.

The inhibitory stimuli, whether nervous or hormonal, are initiated by the presence of various substances in the duodenum, some, such as fatty acids, having a very marked effect while others, such as glucose or mineral salts, have far less. Hunt & Knox (1968) suggested the presence of three types of receptor in the duodenal mucosa. One of these, the least sensitive, was non-specific and responsive to the effect on osmolality of the substance in question, and was responsible for the effects of sugars and the majority of salts on gastric emptying. The second receptor, sensitive to acids with a pH below 5, to a degree inversely related to their molecular weight, was responsible for the effect of acids other than higher fatty acids. The third and most sensitive was responsive to fatty acids with twelve or more carbon atoms.

Within the first group, the effects of some substances are greater or less than that predictable from their osmotic effect in the lumen, and may be related to the ease with which they pass into the mucosa or the receptor to exert their effect. For example, glucose and galactose have more effect than fructose (Elias, Gibson, Greenwood, Hunt & Tripp, 1968) or sorbitol (Reed & Kidder, 1972), while tryptophan has more effect than other amino acids (Stephens, Woolson & Cooke, 1975). Hunt & Stubbs (1975) showed that the overall effect was that the nutritive density (Kcal/ml) of a meal determined the rate of emptying (ml/min). The higher the nutritive density, the slower the emptying, the effect being the same whether the nutrient was carbohydrate, protein or fat.

Studies on gastric emptying in the pig which have been reviewed by Laplace (1975) and Rérat, Corring & Laplace (1976), indicate that, in general, the process is controlled by the same factors as in other mammals.

The actual process of emptying is effected by the contractions which pass down the pyloric region of the stomach into the duodenum (Neimeier, 1940), and it is the motility of the pyloric region which directly influences the emptying after a meal (Auffrey, Martinet & Rérat, 1967).

Of the factors influencing this motility, pH in the duodenum has a relatively slight effect (Auffray, Martinet & Rérat, 1967; Laplace, 1974b; Auffray, 1975). On the other hand, an inhibitory effect on gastric emptying is produced by food in the first eight metres of small intestines in the pig (Auffray, 1975), whereas in many species, only food in the duodenum has an effect.

The approximately exponential emptying after a small feed, or initially after a large feed is also observed in the pig. Reed & Kidder (1972) showed that gastric emptying of 5% glucose solutions by 2–5 month old pigs was approximately exponential with time and that the rate of emptying decreased with increasing glucose concentration. Ochia (1973), with pigs of a similar age, measuring the emptying of a sucrose pectin meal, found that the rate of emptying varied directly with the volume of the meal introduced into the stomach. Laplace & Tomassone (1970), with 70–80 kg pigs with re-entrant duodenal fistulas, which were fed a gruel, found that the emptying could not be reduced to a simple mathematical function, but could be expressed as a polynomial where the coefficients were related to the previous and present state of fullness of the stomach and the small intestine.

Kvasnitskii (1951) showed, in pigs with re-entrant duodenal fistulas, that passage of stomach contents into the duodenum continued throughout the 24 hours with intermittent intensity, being greatest in the 3 hours after feeding. He found considerable individual variation, and also variation in emptying with the quantity of food given, so that 10, 16 or even 24 hours after a large feed, much more remained in the stomach than the same time after a smaller feed. There was also some dependence on the time of day, emptying being slower during the night. This was illustrated by results on pigs given equal feeds at 8-hourly intervals where 61% of the dry matter leaving the stomach during the 24 hours passed in the 8 hours after the early morning feed, and only 16% in the 8 hours after the late evening feed. Thus two major factors stimulating emptying were fullness of the stomach and activity of the animal.

Auffrey, Martinet & Rérat (1967) using the same technique, supplemented by balloon studies of pressure, recorded quantities passed each 15 minutes. They found that during the initial 15 minutes after a meal, some 40% of the stomach contents passed into the duodenum, the volume emptied over this period being approximately proportional to that ingested. Then followed a period when emptying was inhibited, followed by intermittent emptying over the remaining seven hour period studied (Fig. 2.3). Freeman, Noakes, Annison & Hill (1968), also using re-entrant duodenal fistulas, and recording quantities hourly, obtained a high initial rate of emptying which fell for 2–3 hours and then rose and fell again. Measurements of portal blood sugar levels after carbohydrate feeds, described in Chapter 6, (e.g. Aumaitre, Rérat, Vaissade & Vaugelade, 1973) have indicated rhythmical, intermittent emptying into the absorbing region (the proximal small intestine). Where a large feed was given, the initial emptying was greatly increased. The intermittent emptying of the stomach of the pig when it has been fed at a conventional level, is also evident from the studies on pigs with duodenal fistulas reported by Horszczaruk (1971), Zebrowska (1973), Zebrowska, Buraczewska, Buraczewski & Horszczaruk (1975) and Braude, Fulford & Low (1976) and described later in this chapter under the heading "rate of flow".

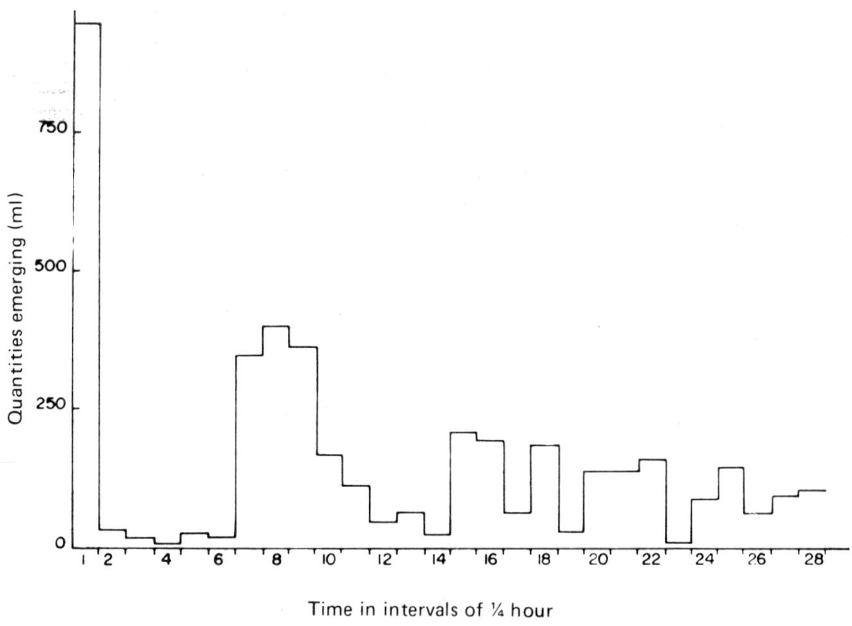

Fig. 2.3 Example of the curve showing the emergence of a meal through a re-entrant duodenal fistula. (Auffray, Martinet & Rérat, 1967)

The way in which the nature of a meal can determine whether the emptying is continuous or intermittent was shown by Reed & Kidder (1971) who found intermittent emptying when pigs were given 50% glucose, but continuous emptying with weaker solutions. Auffray, Martinet & Rérat (1967) and Auffray (1975) showed that replacing starch in a diet by glucose caused the emptying pattern to become more obviously discontinuous (Fig. 2.4). The free glucose clearly produced a more rapid and abrupt response than glucose released gradually from starch. The effect of diet on the initial emptying of a meal was shown by Zebrowska & Buraczewska (1972) in pigs with re-entrant duodenal fistulas, where a protein-free meal was very much more completely emptied from the stomach in the first half-hour than comparable meals containing protein.

Many studies on the pig and other species have not shown this intermittent emptying, but this is probably due to technique, as a small feed will only give the initial approximately exponential emptying. In a slaughter experiment in which pigs were killed at various times after a normal feed, Rérat & Lougnon (1963) obtained an apparently exponential curve. As only one measurement would be made on each pig, it is possible that the emptying curves on the individual pigs were intermittent but that taking one point from each pig gave an approximately exponential curve.

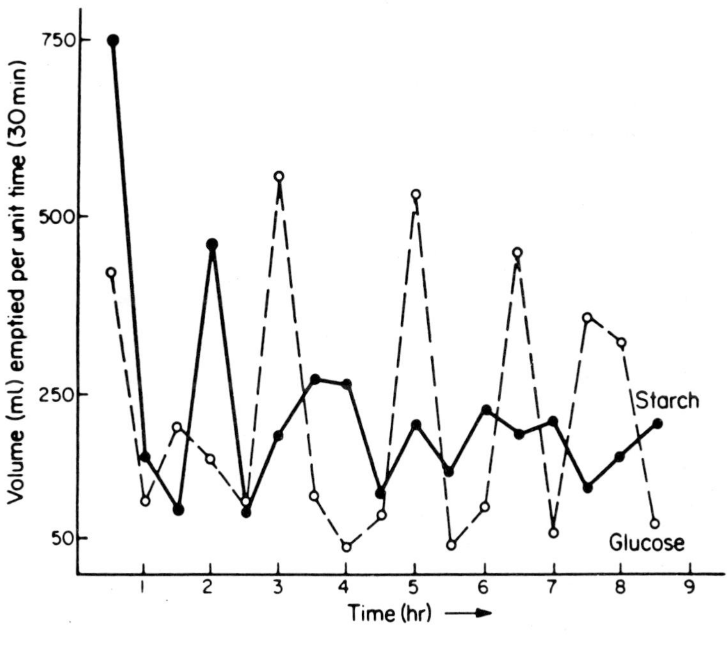

Fig. 2.4 *Gastric emptying of a meal containing 65 per cent carbohydrate (starch or glucose). Digesta collected at a proximal duodenal cannula. The intermittent emptying on the glucose diet shows the inhibitory effect of glucose on gastric emptying (from Auffray, 1965)*

Duration of stay in the stomach

Horszczaruk (1962) on the basis of rubber bead experiments, deduced that food could spend up to 33 hours in the stomach of the pig. This is in agreement with the finding of Neimeier (1940) and Kvasnitskii (1951) that the stomach of the mature pig did not fully empty even after a fast of 24 hours or more.

Kratinov and Kratinova (1932) found that if the pig was fasted long enough for the stomach to become empty, a situation which does not occur in normal management, there was regurgitation of bile and pancreatic juice into the stomach. Ruckebusch & Bueno (1976) reported occasional regurgitation of bile but did not indicate the circumstances under which it occurred. Reed and Kidder (1970) suggested that duodenal regurgitation could be a cause of gastric ulceration in pigs.

Kvasnitskii (1951), on the basis of fistula and slaughter studies, reported that, in the suckling piglet, most of the contents of the stomach emptied between one hourly feeding and the next, an observation supported by Kidder & Manners (1968) on radiographic evidence. In artificially-reared piglets on a twice-daily feeding regime, Kidder, Manners & McCrea (1961) found very much longer emptying times.

THE SMALL INTESTINE

Neimeier (1940), on the basis of fluoroscopic observations, gave times of passage in the small intestine for cellulose-containing materials ranging from two to six hours, with a usual passage time of two and a half to three hours. Kidder & Manners (1974) reported passage times for the small intestine of about two and a half hours in the piglet.

THE LARGE INTESTINE

Horszczaruk (1962), using pigs with caecal fistulas and rubber pellets as markers, measured the time of passage along the large intestine. He found the time interval from the insertion of the marker through the caecal fistula to its first appearance in the excreta to be 12–18 hours and to its disappearance again from the excreta to be 48–54 hours.

Hecker & Grovum (1975), in caecally fistularized pigs, calculated the time of passage through the large intestine to be approximately 30 hours, and the rate of movement to be about 0.2 cm per minute in the proximal 60% of the large intestine. It rose progressively in the distal part to 0.5 to 0.9 cm per minute in the distal 20% of the length.

DISTRIBUTION OF A MARKED MEAL

A method of study giving slightly different information is slaughter of pigs at specified times after a marked meal. Clemens, Stevens and Southworth (1975) slaughtered pigs at different times after a meal containing solid markers of three sizes, and a liquid marker. The results showed the likely distribution of different components of a feed at various times after ingestion (Figs. 2.5 & 2.6).

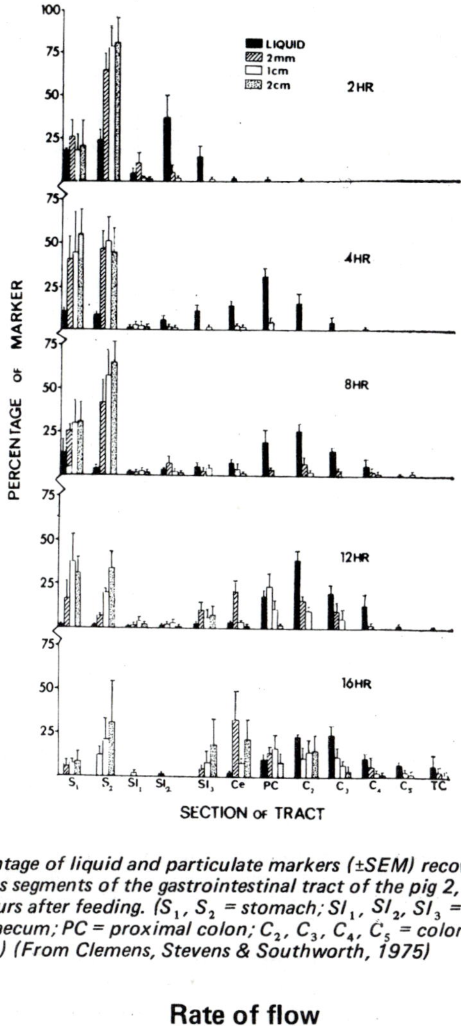

Fig. 2.5 Percentage of liquid and particulate markers (±SEM) recovered from the various segments of the gastrointestinal tract of the pig 2, 4, 8, 12 and 16 hours after feeding. (S_1, S_2 = stomach; SI_1, SI_2, SI_3 = small intestine; Ce - caecum; PC = proximal colon; C_2, C_3, C_4, C_5 = colon; T.C. = terminal colon.) (From Clemens, Stevens & Southworth, 1975)

Rate of flow

A number of authors have reported studies of flow of digesta through the gastrointestinal tract of pigs with re-entrant fistulas. Where these studies have been made using diets similar to those used in agricultural practice, the flow rates may give an indication of the amount of material passing through given parts of the tract in unit time under practical conditions. Studies on gastric emptying, made on pigs with re-entrant duodenal fistulas, have already been mentioned. Buraczewski, Chamberlain, Horszczaruk & Zebrowska (1972) measured the flow rates of total digesta and dry matter through a loop in the proximal jejunum of a 90 kg pig, and showed that this fell rapidly in the second hour after feeding, and then more slowly.

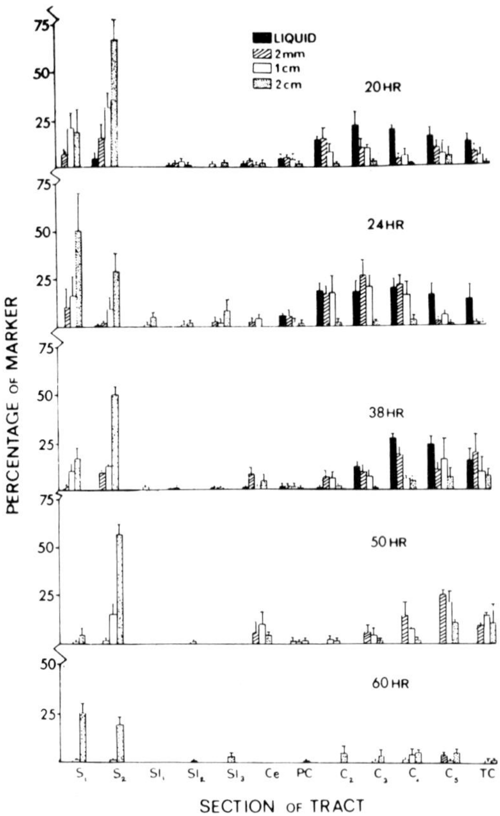

Fig. 2.6 Percentage of liquid and particulate markers (±SEM) recovered from the various segments of the gastrointestinal tract of the pig 20, 24, 38, 50 and 60 hours after feeding. (Symbols as in Fig. 2.5) (From Clemens, Stevens & Southworth, 1975)

Zebrowska & Horszczaruk (1975) found that a once-daily feeding regime in comparison with twice-daily feeding altered the pattern of flow through duodenal and ileal re-entrant cannulas, but did not alter the total volume flowing through either cannula over the course of 24 hours.

Zebrowska (1973) and Zebrowska, Buraczewska, Buraczewski & Horszczaruk (1975) measured flow through duodenal and ileal cannulas on 9 pigs on several diets, with, on the whole, similar findings. The flow rate varied much more in the duodenum than in the ileum, and was greatest in the first hour or two after feeding and least after the overnight fast. The dry matter flow in the duodenum was similar to that in the food intake, but by the lower ileum, this had fallen to one third to one fifth. The total digesta flow, including fluid, in the duodenum was two to three times the intake and, by the lower ileum, was about half the intake.

In similar experiments Cuperlović, Hristić & Zebrowska (1975) noted the wide variation in flow rate at different times after feeding at the duodenal fistula, smaller differences at the jejunal fistula, and an almost constant flow at the ileal fistula.

Braude, Fulford & Low (1976) measured the flow of digesta in the duodenum, the jejunum and the distal ileum of 17–19 kg pigs with re-entrant cannulas. The pigs were fed on three different diets (barley meal, wheat offal, fish meal; starch, sucrose, groundnut meal; starch, sucrose, casein) and the flow rates were expressed as a fraction of the total 24 hourly intake of food and water. Their results for the barley meal, wheat offal, fish meal diet (Fig. 2.7)

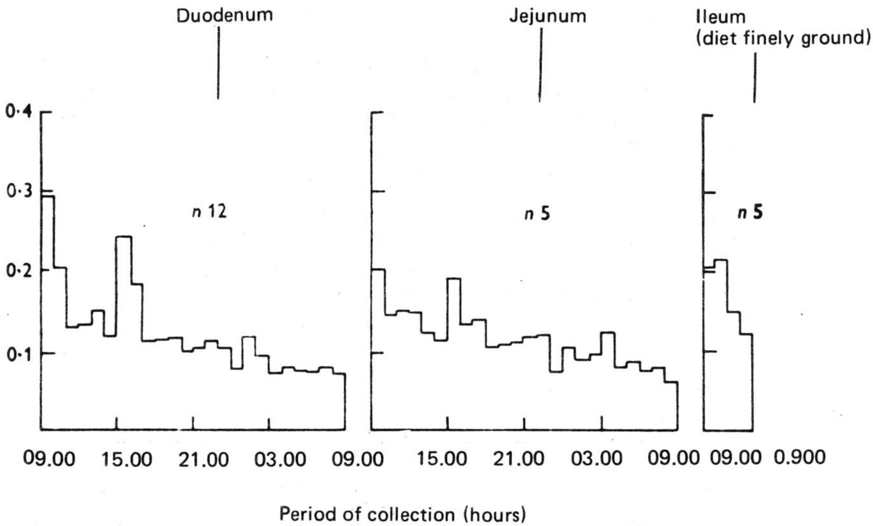

Fig. 2.7 Mean hourly flow (6 h flow in ileum) of digesta (digesta output: feed and water intake) (output: intake) in pigs each fitted with a re-entrant cannula at one of three sites and given a barley meal, wheat offal, fish meal diet. Pigs were fed at 09.00 and 15.00 hours; n is the number of pigs completing collections. (Braude, Fulford & Low, 1976)

show the high flow rate in the duodenum and jejunum after feeding, which fell very sharply in the next 2–3 hours, and then slowly and irregularly over the rest of the 24 hours. Horszczaruk (1971), in similar studies on the large intestine, found only relatively small decreases in dry matter flow along the tract but large decreases in total digesta flow. On a mean daily dry matter intake of 3.2 kg,

Horszczaruk found daily dry matter flows of 0.9 kg in the caecum, 0.73 kg in the colon and 0.63 kg in the faeces. The total digesta flows at the same time were 9.75 kg in the caecum, 5.3 kg in the colon and 2.24 kg in the faeces. This confirms that the predominant process in the large intestine is water absorption.

Ivan & Farrell (1976), comparing hard wheat, soft wheat and semipurified diets on pigs with duodenal and ileal fistulas, also found the flow rate to be far more variable in the duodenal fistula, where it was maximal in the first hour after feeding. The maximum flow rate at the ileal fistula varied with the diet being 9 hours after the feed with soft wheat, but earlier on the other diets, a difference which was attributed to the different degree of mixing of the diet with secretions in the stomach.

Mixing of Food

According to most of the studies on overall time of food passage, a meal first reaches the anus 10–24 hours after ingestion and then continues to emerge until 48–90 hours after ingestion. This means that an individual meal emerges over the course of 1–3 days, although the animal is fed more than once daily. Thus, considerable mixing of the meal with previous and subsequent feeds must occur.

Gastric mixing

The extent of mixing of food in the stomach by gastric movement was studied by Kvasnitskii (1951) who stained successive meals different colours and then slaughtered the pigs, froze the stomachs and cut sections through the contents in various directions. He found little mixing, the layers remaining sharply defined throughout gastric digestion unless liquid mashes were fed, when partial

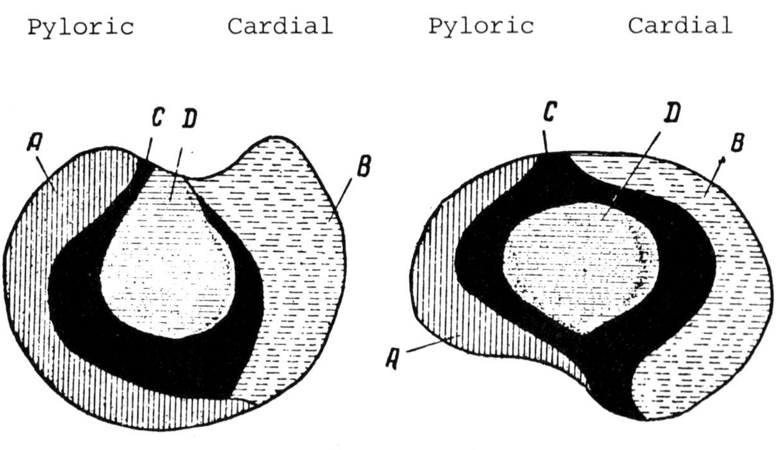

Fig. 2.8 Longitudinal section (vertical, left; horizontal, right) of a frozen stomach of a pig killed immediately after feeding. A – maize stained with yellow dye, B – unstained maize, C – maize stained with blood and D – conventional cereal food mixture. (From Kvasnitskii, 1951)

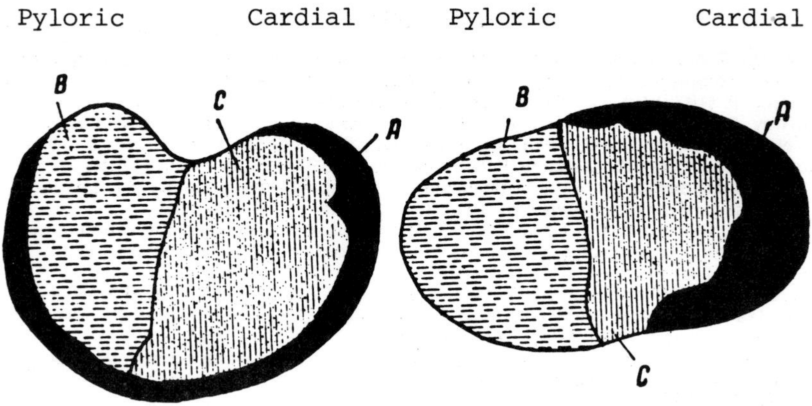

Fig. 2.9 Longitudinal section (vertical, left; horizontal, right) of a frozen stomach. The animal was killed after the evening feed. The food had not mixed. A – maize stained with blood, B – unstained maize and C – maize stained with yellow dye. (From Kvasnitskii, 1951)

mixing occurred. Figs. 2.8 & 2.9 show horizontal and vertical sections of stomachs he prepared in this way from pigs slaughtered after the morning and after the evening feed. Maxwell, Reimann, Hoekstra, Kowalczyk, Benevenga & Grummer (1970, 1972) also using a frozen-stomach technique, showed that finely ground cereal foods mixed much more in the stomach than coarsely ground feeds, and regarded this as a possible cause of gastric ulceration in pigs on finely ground cereal diets.

It does not follow that the layers of food pass on from the stomach exactly in the sequence of ingestion. As was noted by Neimeier (1940, 1941), a considerably amount of food residue is present from previous feeds at the time of feeding a marked meal, yet the marked food started to pass into the duodenum within a few minutes of ingestion. Considerable mixing of different meals must therefore occur during the passage of the food from the stomach into the intestines.

As already mentioned, Ivan & Farrell (1976) attributed differential passage of food on hard wheat and soft wheat diets to different degrees of mixing of food with secretions in the stomach.

Intestinal mixing

Little specific study has been made in the pig of the degree of mixing of successive feeds which occurs in the intestines. It is possible to deduce that this is considerable from the radiographic observations of Kidder, Manners & McCrea (1961) and Kidder & Manners (1968) that with sow-reared pigs, where retention in the stomach is only an hour or two, a marked feed will nevertheless emerge from the rectum over the course of one to three days. With older pigs it would be necessary to insert the marker into the duodenum, otherwise the results would be obscured by the variable time the marked meal spent in the stomach.

UNIVERSITY OF BRISTOL LIBRARY
VETERINARY SCIENCE

REFERENCES

Argenzio, R.A. & Southworth, M. (1974) Sites of organic acid production and absorption in gastrointestinal tract of the pig. *Am. J. Physiol., 228,* 454–460.

Auffray, P. (1975) Cited by Rérat, Corring & Laplace (1976).

Auffray, P., Martinet, J. & Rérat, A. (1967) Quelques aspects du transit gastro-intestinal chez le porc. *Annls Biol. anim. Bioch. Biophys., 7,* 261–279.

Aumaitre, A., Rérat, A., Vassaide, P. & Vaugelade, P. (1973) Étude expérimentale qualitative de l'absorption intestinale des glucides après ingestion d'un repas a bas de glucose ou d'amidon. *Annls Biol. anim. bioch. biophys., 13,* 784–788.

Braude, R., Fulford, R.J. & Low, A.G. (1976) Studies on digestion and absorption in the intestines of growing pigs. Measurements of the flow of digesta and pH. *Br. J. Nutr., 36,* 497–510.

Buraczewski, S., Chamberlain, A.G., Horszczaruk, F. & Zebrowska, T. (1972) Trawienie bialka i wchlanianie lizyny i argininy w izolowanej petlie jelita czcsego świni. *Zesz. Problem. postep. Nauk roln., 126,* 135–140.

Carillo, O. & Benevides, M.C. (1971) Digestibility and rate of passage through the gastro-intestinal tract of torula/molasses in pig diets. *Rev. cubana Cienc. agric. (Eng. ed.), 5,* 331–339.

Castle, E.J. & Castle, M.E. (1956) The rate of passage of food through the alimentary tract of pigs. *J. Agric. Sci., 47,* 196–204.

Castle, E.J. & Castle, M.E. (1957) Further studies of the rate of passage of food through the alimentary tract of pigs. *J. Agric. Sci., 49,* 106–112.

Clemens, E.T., Stevens, C.E. & Southworth, M. (1975) Sites of organic acid production and pattern of digesta movement in the gastrointestinal tract of swine. *J.. Nutr., 105,* 759–768.

Cooke, A.R. (1975) Control of gastric emptying and motility. *Gastroenterology, 68,* 804–816.

Cuperlović, M., Hristić, V. & Zebrowska, T. (1975) Influence of the composition of the dietary protein supplement on the digestion and absorption of protein in the small intestine of pigs. *Acta Veterinaria, Belgrade, 25,* 287–300.

Elias, E., Gibson, G.J., Greenwood, L.F., Hunt, J.N. & Tripp, J.H. (1968) The slowing of gastric emptying by monosaccharides and disaccharides in test meals. *J. Physiol., Lond., 194,* 317–326.

Falaschini, A. & Carucci, F. (1934) Experienze sui tempi di passagio degli alimente nell' apparato digerente dei suini. *La Nuova Vet., 12,* 397–401.

Fossum, K. & Liven, E. (1974) The distribution of enzymes and bacteria in the small intestinges of slaughter pigs. *Acta path. microbiol. scand., 382,* 644–652.

Freeman, C.P., Noakes, D.E., Annison, E.F. & Hill, K.J. (1968) Quantitative aspects of intestinal fat absorption in young pigs. *Brit. J. Nutr., 22,* 739–749.

Green, D.A. & Jewell, P.A. (1965) A method of measuring food passage rate in pigs by means of polystyrene beads. *Brit. vet. J., 121,* 83–9.

Hahn, W. (1975) Untersuchungen zur Magenmotorik des Mastschweines mittels Radiotelemetrie und elektronischer Datenverarbeitung. Diss. Hannover.

Hecker, J.F. & Grovum, W.L. (1975) Rate of passage of digesta and water absorption along the large intestine of sheep, cows and pigs. *Aust. J. biol. Sci., 28,* 161–167.

Hopkins, A. (1966) The pattern of gastric emptying: a new view of old results. *J. Physiol, Lond., 182,* 144–149.

Horszczaruk, F.. (1962) Wplyw zroznicowanego poziomu wlókna surewego w dawkach na procesy trawienne u świń. II Dlugość i projemność przewodu pokarmowego oraz czas przechodzenia treści pokarmowej. *Roczn. Nauk Roln., 80–B–2,* 115–125.

Horszczaruk, F. (1971) Badania nad trawieniem u świń. I. Badania z zastosowaniem przetok ślepego i okreznicy. *Biul. Inst. Genetiki i Hodowli Zwierzat PAN, 21,* 101–115.

Hunt, J.M. & Know, M.T. (1968) Regulation of gastric emptying. *in* Handbook of Physiology, Sect. 6, vol. 4, pp 1917–1935. American Physiological Society.

Hunt, J.N. & MacDonald, J. (1954) The influence of volume on gastric emptying. *J. Physiol. Lond., 126,* 459–474.

Hunt, J.N. & Stubbs, D.F. (1975) The volume and energy content of meals as determinants of gastric emptying. *J. Physiol., Lond., 245,* 209–225.

Ivan, M. & Farrell, D.J. (1976) Flow rates of components in digesta of pigs prepared with re-entrant cannulas in the proximal duodenum and terminal ileum, and fed semipurified, hard wheat, and soft wheat diets. *Can. J. Physiol. Pharm., 54,* 891–897.

Keys, J.E. & DeBarthe, J.V. (1974) Site and extent of carbohydrate, dry matter, energy and protein digestion and the rate of passage of grain diets in swine. *J. Anim. Sci., 39,* 57–62.
Kidder, D.E. & Manners, M.J. (1968) Passage of food in the young suckling pig. *Proc. Nutr., Soc., 27,* 46A.
Kidder, D.E. & Manners, M.J. (1974) Passage of food in the piglet. 3rd Intern. Congr. (IPVS) Lyon. L11, 1–3.
Kidder, D.E., Manners, M.J. & McCrea, M.R. (1961) The passage of food through the alimentary tract of the piglet. *Res. vet. Sci., 2,* 227–231.
Kratinov, A.G. & Kratinova, P.N. (1932) Materialy k fiziologii motornoi deyatel'nosti pishchevaritel'nogo trakta svin'i. *Sechenov physiol J. U.S.S.R., 15,* 492–500.
Krüger, L. & Meyer, F. (1958) Untersuchungen über die Durchgangszeiten von Futtermitteln durch den Verdauungskanal bei Schweinen. *Züchtungskunde 7,* 324–331.
Kvasnitskii, A.V. (1951) Voprosy fiziologii pishchevareniyau svinei. Sel'khozgiz, Moscow.
Laplace, J.P. (1972a) Motricité gastro-intestinale chez le porc; étude descriptive par électromyographie et corrélations nutritionelles. *Recl. Méd. Vét., 148,* 37–61.
Laplace, J.P. (1972b) Existence d'un rhythm électrique de type antral au niveau des voies biliaires du porc. *Annls Biol. anim. Biochim. Biophys., 13,* 708–710.
Laplace, J.P. (1974a) Activité électrique lente de l'antre pylorique et contrôle de l'évacuation gastrique chez le porc. *Recl. Méd. vét., 150,* 121–129.
Laplace, J.P. (1974b) Enregistrement du pH intragastrique chez le porc: variations lieées à la nature, à l'importance et à l'intervalle des repas chez l'animal en finition. *Annls Zootech., 23,* 89–104.
Laplace, J.P. (1975) Digestive transit in monogastric animals. Review 2. Motor events and digesta movements. *Annls. Zootech., 24,* 489–551.
Laplace, J.P. & Tomassone, R. (1970) Évacuation gastro-duodénale chez le porc. Fistulation chronique par voie thoracique extrapleurale; recherche d'une technique d'analyse mathématique d'évacuation. *Annls Zootech., 19,* 303–332.
Maner, J.H., Pond, W.G., Loosli, J.K. & Lowrey, R.S. (1962) The effect of isolated soyabean protein and casein on the gastric pH and rate of passage of food residues in baby pigs. *J. Anim. Sci., 21,* 49–52.
Marsboom, R., Temmerman, R. & Symoens, J. (1973) The effect of benzetimide on intestinal motility in pigs. *Vet. Rec., 93,* 382–384.
Maxwell, C.V., Reimann, E.M., Hoekstra, W.G., Kowalczyk, T., Benevenga, N.J. & Grummer, R.H. (1970) Effect of dietary particle size on lesion development and on the contents of various regions of the swine stomach. *J. Anim. Sci., 30,* 911–922.
Maxwell, C.V., Reimann, E.M., Hoekstra, W.G., Kowalczyk, T., Benevenga, N.J. & Grummer, R.H.. (1972) The use of tritiated water to assess in vivo, the effect of dietary particle size on the mixing of stomach contents of swine. *J. Anim. Sci., 34,* 212–216.
Neimeier, K. (1940) Röntgenologische Beobachtungen am Magen–Darmkanal des Schweines. Diss. Hannover.
Neimeier, K. (1941) Roentgenologische Beobachtungen an Magen Darmkanal des Schweines. *Jahresb. Vet. Med., 68,* 6–8.
Ochia, B.A. (1973) Gastric emptying in young pigs. *J. Physiol., Lond., 233,* 467–480.
Plonait, H. (1974) Measurement of stomach motility using surgically implanted telemetry transmitters. 3rd Intern. Congr. (IPVS) Lyon. L 10 1–7.
Reed, J.H. & Kidder, D.E. (1970) A post mortem survey of the relationships between bile staining and ulceration of the oesophageal portion of pig's stomach. *Res. vet. Sci., 11,* 438–440.
Reed, J.H. & Kidder, D.E. (1971) The oral glucose tolerance test in the young pig. *Br. vet. J., 127,* 318–326.
Reed, J.H. & Kidder, D.E. (1972) The effect of glucose, galactose and sorbitol on gastric emptying in the pig. *Q. Jl. exp. Physiol., 57,* 30–36.
Rérat, A., Corring, T. & Laplace, J.P. (1976) Protein digestion and absorption. *in* Protein metabolism and nutrition. Eds Cole, D.J.A., Boorman, K.N., Buttery, P.J., Lewis, D., Neal, R.J. & Swan, H. pp. 97–138. Butterworth, London.
Rérat, A. & Lougnon, J. (1963) Études sur le transit digestif chez le porc. *Annls Biol. Anim. Biochim. Biophys., 3,* 21–29.
Ruckebusch, Y. & Bueno, L. (1976) The effect of feeding on the motility of the stomach and small intestine in the pig. *Br. J. Nutr., 35,* 397–405.

Seerly, R.W., Miller, E.R. & Hoefer, J.A. (1962) Rate of food passage studies with pigs equally and ad libitum fed meal and pellets. *J. Anim. Sci., 21,* 834–7.

Starovoitov, A.M. (1960) Age changes in the motor activity of the stomach of young pigs. *Sechenov physiol. J. U.S.S.R., 46,* 572–578.

Stevens, J.A. & Kidder, D.E. (1972) The distributionof trehalase, sucrase, α-amylase, glucoamylase and lactase (β-galactosidase) along the small intestine of five pigs. *Br. J. Nutr., 28,* 129–137.

Stephens, J.R., Woolson, R.F. & Cooke, A.R. (1975) Effects of essential and nonessential amino acids on gastric emptying in the dog. *Gastroenterology, 69,* 920–927.

Thomas, J.E. (1957) Mechanics and regulation of gastric emptying. *Physiol. Rev., 37,* 453–474.

Thurber, E.T., Hudman, D.B. & Peo, E.R. (1965) Rate of food passage in swine. *J. Anim. Sci., 24,* 930 (abs. 353)

Trautmann, A. & Ascher, T. (1940) Röntgenologische Untersuchungen zur Motorik des Colons des Schweines. *Archs. wiss. prakt. Tierheilk., 76,* 119–132.

Zebrowska, T. (1973) Influence of dietary protein source on the rate of digestion in the small intestine of pigs. I. Amount and composition of digesta. *Roczn. Nauk Roln., 95–B–1,* 115–133.

Zebrowska, T. & Buraczewska, L. (1972) Wplyw poziomu bialka w diecie na przebieg trawienia w jelicie cienkim świń. I. Ilość i sklad treści. *Roczn. Nauk Roln., 94–B–1,* 81–96.

Zebrowska, T., Buraczewska, L., Buraczewski, S. & Horszczaruk, F., (1975) Digestion and absorption in the small intestine of pigs. I. Digestion and absorption of dry matter and nitrogen. *Roczn. Nauk Roln., 96–B–3,* 79–90.

Zebrowska, T.. & Horszczaruk, F. (1975) Wplyw jedno- i dwukrotnego karmenia na ilość i sklad treści jelita cienkiego świń. *Roczn. Nauk Roln., 96–B–3,* 91–96.

CHAPTER 3
CONDITIONS IN THE LUMEN. pH.

CONSISTENCY AND COMPOSITION

While the consistency and composition of the contents of the gastrointestinal tract will be affected very considerably, especially in the stomach, by the nature of the diet, the composition in any particular section of the gut will only vary within certain limits on any normal diet. There are a number of studies which give an indication of the sort of composition which could be expected in a normal healthy pig.

Dry Matter Content and Consistency

The consistency of the contents is very largely a reflection of the dry matter content. This is less true in the stomach, where particle size of the diet has a considerable effect, and where a fibrous diet may give a fairly stiff mass, while a finely ground meal will give more fluid contents.

In the duodenum the contents are always fluid, and become progressively thicker during their passage along the tract and especially when reaching the large intestine. Kvasnitskii (1951), Stevens and Kidder (1972) and Fossum & Liven (1974) have drawn attention to the alternation in the small intestine of relaxed or dilated regions filled with partly digested food, and constricted regions where the lumen is almost closed. Fossum & Liven noticed that in the relaxed regions, the consistency of the contents varied from being pasty or "porridge-like" to being watery or "soup-like", while, in the constricted regions, it was always "mayonnaise-like".

Figures for the dry matter content along the gastrointestinal tract were given by Kvasnitskii (1951) on pigs on different diets slaughtered at various times after feeding. Varying the time after feeding between 5 and 15 hours made no significant difference. The level of protein in the diet only influenced the dry matter levels in the stomach, where the mean values were 25–26% on a low protein diet and 14–20% on a high protein diet. The dry matter concentrations for the proximal and distal halves of the small intestine, the caecum and the large intestine were 6–9%, 9–12%, 10–16% and 13–24% respectively. In studies on the gut contents of 26 slaughtered pigs, Horszczaruk (1971a) obtained dry matter values of 22.8 ± 4.65% for the stomach and 13.38 ± 2.22% and 13.40 ± 0.80% for the first and second half of the small intestine respectively. Percentage moisture in gastric contents of pigs was also measured by Maxwell, Reimann, Hoekstra, Kowalczcyk, Benevenga & Grummer (1970). They reported levels from just above 60% to nearly 100% depending on time after feeding and the ration. The moisture rose progressively after feeding, and was higher throughout, and rose more rapidly on a finely ground maize ration than on a coarse ration.

Levels of dry matter in the digesta have been measured on pigs with re-entrant fistulas, where it was only possible to sample at certain points but samples could be taken over a considerable period of time. Horszczaruk (1971b), in experiments on pigs with re-entrant fistulas just below the pylorus and about half way along the small intestine, found the mean dry matter content of digesta collected over a period of hours to be between 8.5% and 15%. Only on a high protein diet was the dry matter appreciably lower in the duodenal fistula than in the more distal one. Sampling contents from the duodenal fistula at different times after feeding, he found that the dry matter rose slightly from 10.6% in the first half hour to over 12% in the second half hour. It then fell to 6–9% for the next 3 hours and then to 4–6%.

In a number of other studies on fistularized pigs, the percentage dry matter has not been given but can be calculated from the data (Horszcsaruk, 1971c; Zebrowska & Buraczewska, 1972; Zebrowska, 1973; Buraczewska, Buraczewski, Raczyński & Zebrowska, 1973). The most extensive data are those of Zebrowska (1973) who found that, on a range of diets, the dry matter content in the duodenum varied from 15–20% shortly after feeding to 2.5–5% after an overnight fast, while in the distal ileum the dry matter was in the range of 6–13% with relatively little variation with time after feeding.

Keys & DeBarthe (1974), in pigs on diets with high levels of hay, obtained dry matter values of 4.8–6.85% in the duodenum, 4.8–6.4% in the distal ileum and 24–35% in the faeces.

In the contents of the large intestine of the pig, Hecker & Grovum (1975) showed that the dry matter rose sharply from 13% at the caecum to 20% in the proximal third of the length, and then more slowly, barely reaching 25% dry matter by the rectum, in contrast to the sheep and cow, where the rise in dry matter was more continuous and in the sheep, much greater (Fig. 3.1).

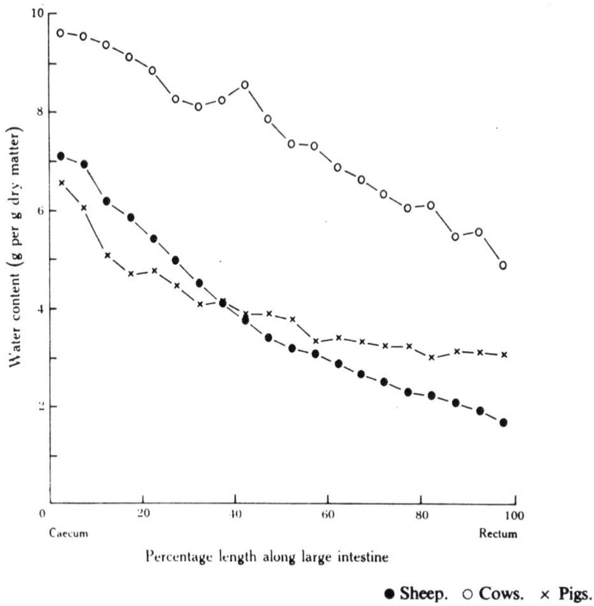

Fig. 3.1. Water content of digesta along the large intestine (Hecker & Grovum, 1975)

Osmolality

There has been little study of the osmotic concentration of pig gut contents, though this must be of considerable importance in influencing the rate of absorption of solutes, and especially the movement of water between the lumen and the mucosa. Maxwell, Reimann, Hoekstra, Kowalczyk, Benevenga & Grummer (1970) found values of 300–700 mOsM for the stomach contents of pigs, the value depending on the portion of the stomach, the time after feeding and the nature of the food. With finely ground feeds where gastric mixing occurred, the levels were similar in all sections of the stomach, falling from 500–600 mOsM just after feeding to about 300 mOsM after 8 hours. With coarse foods, osmolality fell similarly in the pyloric region of the stomach, but rose slightly in the oesophageal portion, and fell to low values after more than 13 hours.

In the intestines the osmolality shows less variation. Harpur & Popkin (1965) measured the osmolality of the small intestine contents of pigs at slaughter, and obtained values just below 400 mOsM where the pigs had had food and water before slaughter and 427 when they had been fasted for 24 hours before slaughter.

REACTION

One of the most noteworthy features of the process of digestion of food by mammals is the way in which digestive secretions change the reaction of the digesta during its passage. After being acted upon by alkaline salivary secretions, the food next passes to the stomach where the reaction in older pigs is strongly acid as a result of the secretions of the glands of the fundal region. From the stomach the digesta pass into the duodenum and small intestine where, as described later, alkaline pancreatic juice, bile and the secretions of the

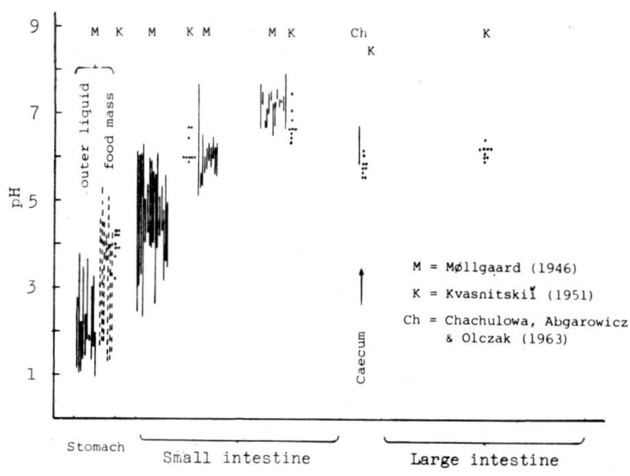

Fig. 3.2 pH in the gut of adult pigs (Manners, 1970)

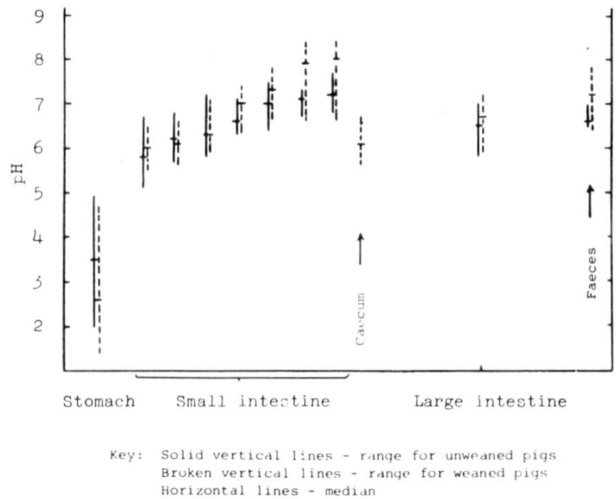

Fig. 3.3 pH in the gut and faeces of unweaned and weaned pigs (Manners, 1970 – Drawn from the data of Smith & Jones, 1963)

intestinal wall change the reaction once again towards neutrality. These secretions are described in Chapter 5. Here we shall deal with measurements of the resulting pH of the contents of the tract.

Values of pH in the gut lumen of mature pigs reported by various earlier authors were summarized by Manners (1970) (Fig. 3.2) Values for suckling and weaned piglets, obtained by Smith & Jones (1963) (Fig. 3.3) are very similar, as are the values given by Argenzio & Southworth (1974) for growing pigs and Clemens, Stevens & Southworth (1975) for mature pigs. The latter two reports show the effect of the time interval after feeding (Fig. 3.4), which has a marked influence on the pH in the stomach and proximal small intestine, but not in the rest of the gut. Figs. 3.2, 3.3 and 3.4 illustrate the difference between the stomach (and to a lesser extent the duodenum), where the pH is dominated by the gastric acid secretion, and the rest of the gut. We shall deal with these two regions separately.

Gastric pH

There is a considerable difference between the piglet, where gastric secretion is extremely limited, and the pig of beyond 8 weeks of age where gastric secretion is well developed, and which will be considered, though perhaps not quite accurately, as a mature pig.

Some caution must be observed in interpreting data on gastric pH, in that some of the measurements are on mixed samples while others have been made at a particular position, usually the base of the stomach, where, as is explained later, the pH is likely to be lower.

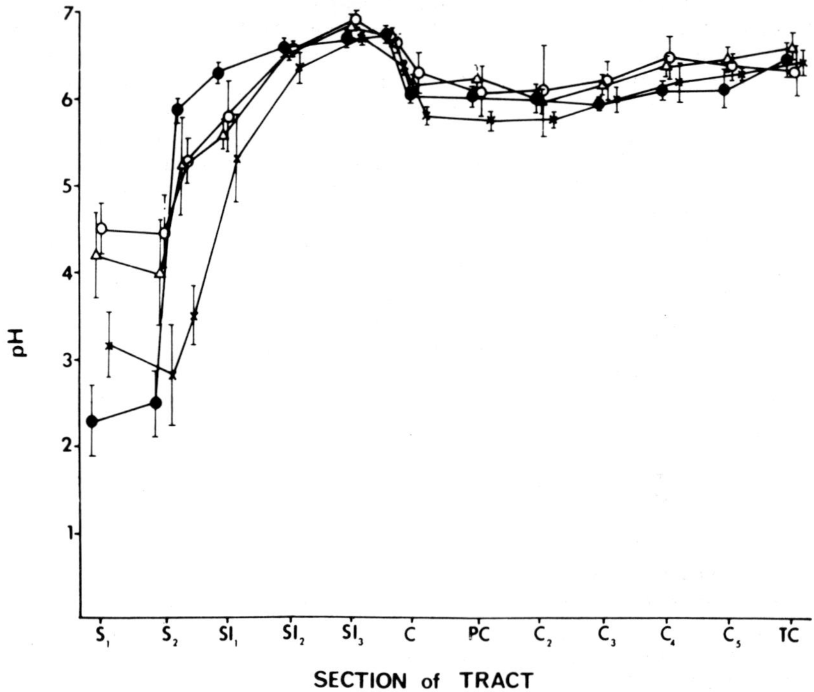

Symbols within the graph correspond to the four time periods: (●), 0 or 12 hours; (△), 2 hours; (○), 4 hours; and (×), 8 hours after feeding. Symbols along the abscissa represent the sections of the tract as given in figure 2.5

Fig. 3.4 Mean pH (±SEM) of gastrointestinal contents of the pig (Clemens, Stevens & Southworth, 1975)

Mature pigs

Mollgaard (1946) showed clearly the difference between the pH of the outer liquid in the stomach and that of the inner food mass. Kvasnitskii (1951) also showed that the acidity in the stomach varied considerably from one part to another. He fitted a triple lumen sampling tube into the stomach of pigs with a fundal fistula, enabling him to sample at three different levels, and sampled the contents at intervals after feeding. An hour after feeding, acidity was only high in the lowest layer, but after seven hours the acid had penetrated to the uppermost layer, the rate of penetration being very dependent on the nature of the food. Slivitskii (1975) used the same fistula technique with 8–10 month old pigs on a starch diet, and obtained pH values of about 1.0 in the basal layer, while the uppermost layer had a pH value of about 3.5 two hours after feeding, falling to about 2.0 twelve to fourteen hours after the feed. He also measured volatile fatty acids and lactic acid, (Chapter 6. Table 6.5). Thus the pH of the contents of the stomach is related to the time after feeding, the nature of the diet and the position at which it is being measured. It will therefore also depend on the degree to which the contents of the stomach are mixed. This was demonstrated by Maxwell et al. (1970) who found that the pH of the contents of the stomach did not vary greatly from one region to another when the pigs were on a finely ground maize diet in which there was gastric mixing, but that on a coarsely ground diet, the usual differences in gastric pH occurred between one region and another. This effect was suggested by Lawrence (1972) as the reason why he obtained lower pH values on more fibrous diets. Lawrence (1970, 1972) showed that the pH rose immediately after feeding to levels depending on both the diet and the feeding regime, but usually in the range pH 4–5. It then fell over the next 2 to 8 hours to pH 1–2, usually just below pH 2, the rate of fall depending both on the composition and the physical nature of the diet.

Laplace (1974), measuring pH in the antrum adjacent to the pylorus, similarly found rises immediately after feeding and a subsequent fall depending on the diet, but found no difference with different feeding regimes.

Gastric pH in piglets

As is described in more detail in Chapter 5, gastric acid secretion is very limited in the piglet for the first few weeks of life. This limitation affects the gastric pH differently in the suckled and the early-weaned piglet.

SUCKLED PIGLETS

Smith & Jones (1963) reported substantially higher pH values (5.2–5.3) in the stomach of the piglet in the first few hours of life, compared with values of 3–4 which he found a few hours later. The pH ranges he gave (Fig. 3.3) cover the ranges found by most other workers.

A factor affecting the gastric pH of the suckled piglet under 6 weeks of age is the mutual interaction of hydrochloric acid secretion and bacterial lactic fermentation reported by Cranwell, Noakes & Hill (1968, 1976) and described in Chapter 6. This effect results in piglets reared in a "conventional" farm environment having a high gastric lactic acid concentration, while those reared in a "clean" environment have, instead, a higher hydrochloric acid concentration and so a rather lower pH. the pH in both cases was nearly always in the range 3.5 to 6, the higher value being just after suckling.

As explained by Cranwell, Noakes & Hill (1976), the relatively high gastric pH in suckling piglets is not due to an inability to secrete hydrochloric acid but to a relatively limited secretory capacity which is not strongly stimulated by sow's milk (Kvasnitskii, 1951), which itself has considerable buffering power (Kvasnitskii, 1951; Kutas, Szabó & Ribiczeyné–Szabó, 1973).

The studies of Schicketanz & Richter (1967) on suckled pigs from 5 weeks of age show that, by this time, the pattern of gastric pH resembles that of a mature pig with levels of 1–2, rising temporarily to 3–4 after a feed. These measurements, made with an endoradiosonde, will have been from close to the fundal wall, and so have given the lowest pH of any part of the stomach.

EARLY-WEANED PIGLETS

Hartman, Hays, Baker, Neagle and Catron (1961), measuring mixed stomach contents, did not find a difference between early weaned and suckled pigs until 5 weeks of age, when the early weaned pigs had lower pH values (3.9 compared with 5.2). Schulman (1973), on the other hand, found higher pH values in the fundus and pylorus of weaned pigs from 5 weeks, as compared with suckled litter-mates. Polivoda, Derevinśkii & Kudrik (1973), measuring pH directly in the stomach, and also gastric juice obtained through a fistula, found lower pH values in the early weaned pigs on a diet containing 50% dried milk together with cereals than in suckled litter-mates from 20 days of age (Table 3.1). Kutas & Szabó (1974), comparing stomach tube samples from suckled piglets and from piglets weaned at 1 week, obtained slightly lower gastric pH values from the weaned animals over the age range 12–22 days (4.43 ± 0.145 compared with 4.92 ± 0.184).

Table 3.1 *pH in the stomach of suckled piglets and piglets weaned at 10 days (from Polivoda, Derevins'kii & Kudrik, 1973)*

Age (days)	pH (mean ± S.E.M.) Suckled	pH (mean ± S.E.M.) Weaned
10	3.77 ± 0.04	4.19 ± 0.45
20	4.77 ± 0.49	3.42 ± 0.18
30	5.02 ± 0.34	4.29 ± 0.34
40	3.61 ± 0.42	3.24 ± 0.53
50	4.19 ± 0.28	3.79 ± 0.69
60	3.46 ± 0.23	2.84 ± 0.38

In all these comparisons, the gastric pH values both of the suckled and the early-weaned pigs were within the ranges found by other reports for the respective treatment, and various workers have shown that a considerable range of pH values can be found in early-weaned, as in suckled pigs, depending on treatment and method of measurement.

In view of the limited capacity for acid secretion in the very young pig, a very important factor determining the pH of the gastric contents of early weaned pigs is the buffering power of the diet. Pointing this out, Manners (1970) demonstrated the very different buffering capacities of different weaning diets and, in particular, the very large buffering capacity of the high-protein diets usually used for very young pigs. This is one reason for the conflicting results different workers have obtained when comparing pH in the stomach of early weaned and suckled pigs. This dependence of gastric pH on the diet in early-weaned piglets is evident from the results of a number of workers (e.g. Maner, Pond, Loosli & Lowrey, 1962; Manners, 1970; Sève & Laplace, 1975).

A second cause of difference, the method of measurement, was demonstrated by Maner et al. (1962), who obtained higher pH values on stomach contents obtained from piglets by stomach tube or at slaughter than those measured in the fundal part of the stomach of the live animal through a fistula, which on 4 week old piglets, were 3.2 after a feed falling to 1.1 after 4 hours on a casein diet and 5.0 falling to 1.3 on a soybean diet. Low values before a meal (1.5–2.5) were reported by Sève & Laplace (1975) using a similar technique on 3–4½ week old weaned piglets. These values usually rose after feeding to pH

5–7 depending on diet and conditions. On 8 week-old early-weaned piglets, Maner et al. (1962) obtained gastric pH values by this fistula technique similar to those found in a mature pig (1.7–1.8 except just after a feed on one diet when it rose to 2.5).

Intestinal pH

The pH values at any point along the intestines vary far less than the values in the stomach, and there is not the same difference between the piglet and the mature pig (Figs. 3.2 and 3.3). The pH in the duodenum can vary considerably (from about 2 to 6), depending on the time after feeding and whether the stomach is emptying, these fluctuations becoming progressively smaller further down the small intestine. This was remarked on by Braude, Fulford & Low

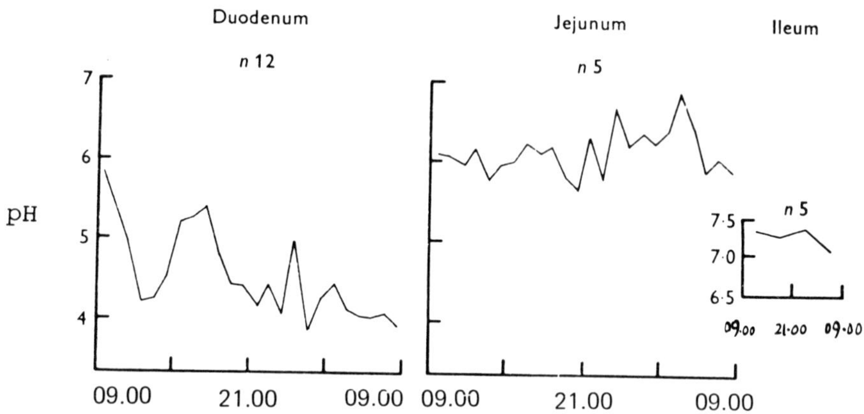

Fig. 3.5 *Mean hourly pH values (6 h value in ileum) in pigs each fitted with a re-entrant cannula at one of three sites. Pigs were fed at 09.00 and 15.00 hours; n is the number of pigs completing collections. (Braude, Fulford & Low, 1976)*

(1976) in their report on pigs with re-entrant cannulas at different points along the small intestine. Fig. 3.5 shows the pH values that they obtained on a barley-wheatfeed-white fish meal diet. The duodenal fluctuations are much greater than the jejunal, in spite of the former being a mean of more animals. The ileal values will have had their fluctuations diminished by being 6-hour rather than 1-hour means. The values obtained by Braude et al., usually in the range pH 4–6 in the duodenum, 5.5–6.7 in the jejunum and 7.0–7.5 in the distal ileum, are similar to those reported by other workers (Figs. 3.2 and 3.3).

The pH in the large intestine has not been reported to fluctuate significantly with time after feeding, and values of around 6.0 have been found by most observers (Fig. 3.2).

In the suckling piglet, Walker (1959) and Smith & Jones (1963) did not find any large change in intestinal pH with age. In the early weaned piglet, Hartman et al. (1961) found lower values than in suckled controls, and Manners (1970) found values in early weaned piglets even lower than those reported by Hartman et al., with values of 5.3 at the proximal end, rising to 5.6 at the distal in 2–3 week-old piglets and 5.7 to 6.4 at 7 weeks. As with gastric pH, it is likely that the intestinal pH of the piglet is appreciably affected by the nature of the diet.

Significance of pH in the Gastrointestinal Tract

Bacterial multiplication

SUCKLED PIGLETS

Smith & Jones (1963), commenting on the higher pH in the stomach contents of suckled piglets on the first day of life than subsequently, observed that the whole alimentary tract was flooded with bacteria in the first day or so of life and that this was probably due to the pH of the food in the stomach at that time being such as to permit rapid multiplication of ingested bacteria, which then passed unharmed into the small intestine and continued to multiply there. The lower pH value of the stomach contents after the first day of life was sufficient to curb the multiplication of all ingested bacteria except the lactobacilli, which continued to proliferate in the stomach of pigs of all ages and formed the principal inhabitants of the upper alimentary tract. The lactic acid production from these lactobacilli can then partially suppress hydrochloric acid production in the suckling pig up to the age of about 5 weeks, as reported in the work of Cranwell, Noakes & Hill (1968; 1976) described in Chapter 6. This will produce conditions favourable to the continued proliferation of lactobacilli.

EARLY-WEANED PIGLETS

Smith & Jones (1963) found that when alkaline substances were added to pig food so that a high pH was maintained while the food was in the stomach, bacteria whose multiplication was normally limited were found to proliferate and pass into the small intestine with very little decrease in numbers.

It would thus seem likely that the small intestine of the piglet reared artificially will be exposed to a constant flow of micro-organisms in the efflux from the stomach until the period is reached when the secretion of hydrochloric acid begins to assume mature proportions. Feeding only twice daily would be likely to aggravate this situation since the rate of stomach emptying is greatest when the stomach is full. Thus, under conditions of deficiency of hydrochloric acid secretion, the greater the amount of food consumed at any one meal, the less likely it is that pH will fall to levels capable of inhibiting bacterial growth, and the more likely it is that a considerable part of the contents will pass from the stomach to the small intestine soon after feeding.

The pH in the stomach of artificially reared piglets can remain high up to 5 weeks of age so there is some justification for presuming that there is little restriction, in these circumstances, on the survival or multiplication of micro-organisms within the stomach contents. Consequently these would move into the lower gut as the stomach emptied.

The studies of Kovács, Nagy & Sinkovics (1972) on the relationship between bacterial flora and local pH in the gut of healthy early-weaned piglets showed the effect of low pH in suppressing organisms other than lactobacilli, but also showed that other factors, such as the rearing environment had as significant an effect.

The ability to rear piglets successfully on artificial diets probably depends to a large extent on factors minimising the loading of the gut, namely care in formulation of diets so that when mixed with water they are as low in bulk as possible, and the use of a limited scale of feeding. Manners (1970) found that if piglets were allowed to eat as much as they wished under twice-daily feeding of liquid diets, after a time certain animals started to scour, then others became ill, and soon there was a major disease problem. It seems likely that feeding small quantities of feed once-hourly would allow the limited capacity of the piglet's stomach to secrete hydrochloric acid to have maximum effect, both on bacterial multiplication and on peptic proteolysis.

Effect on enzymes

Studies of the effect of pH on the activity of digestive enzymes other than pepsin generally show a fairly narrow range of optimum pH and a rapid falling off in rate of hydrolysis of substrate on either side of the peak. The enzymes used in such studies may have gone through involved separation procedures and may no longer show the same pH optima that they would in whatever combination they might exist in the gut of the living animal. Also the type of buffer or the type of substrate used in the assay often affects the pH optimum observed. However, even fairly crude enzyme extracts show peaked pH-activity curves, so whatever the optimum pH for an enzyme may be in vivo, it is fairly certain that, if widely differing pH levels are compared, the activity of the enzyme is likely to be lower at one pH than at the other.

GASTRIC pH

As described in protein digestion, the pepsins all have two pH optima, one near pH 2.0 and the other about pH 3.5. Some proteins such as egg albumen are only digested rapidly at the lower pH. If the pH in the stomach does not fall below 4.0, no significant proteolysis will occur. If it does not approach 2.0, hydrolysis of some proteins will be limited. Kvasnitskii (1951) reported that proteolysis was absent in the stomach of piglets under 3 weeks of age, owing to the lack of acidity.. He found pepsin there but the juice was not sufficiently acid to act on the egg albumen he used as a test material. The work of Cranwell, Noakes & Hill (1968, 1976) indicates that even in piglets reared on a sow under adverse conditions, the higher optimum pH for proteolysis is reached, and that under favourable conditions, the lower pH is reached. Thus, in sow-reared pigs, lack of gastric acidity is less of a limitation on gastric proteolysis than had been supposed.

INTESTINAL pH

The pH in the lumen of the gut will clearly affect the pancreatic enzymes which are secreted into the lumen. It will also affect the enzymes on the brush border of the intestine mucosa to a degree which cannot be accurately assessed as we cannot measure the pH in the spaces between the microvilli of the mucosal cells. Table 3.2 gives the pH optima for pancreatic hydrolases in the pig. In many of the pro-

Table 3.2 Approximate in vitro pH optima for porcine pancreatic hydrolases

Enzyme	Optimum pH	Substrate	Reference
Trypsin	8.2–9.2 7.3–8.6 7.8–8.8	Haemoglobin Gelatin Casein	Northrop et al. (1948)
Chymotrypsin	7.0–9.0 7.8	Proteins Synthetic substrates	Desnuelle (1960)
Chymotrypsin	7.0–9.0	Casein	Northrop et al. (1948)
Carboxypeptidase A	7.0–8.0	Hippuryl-L-phenylalanine	Folk & Schirmer (1963)
Carboxypeptidase B	7.9–8.0	Hippuryl-L-arginine	Wolff et al. (1962)
Elastase	8.8	Elastin	Lewis et al. (1956)
Ribonuclease	7.5	Yeast RNA	Yamasaki et al. (1968)
α-Amylase	6.9	Starch	Fischer & Bernfeld (1948)
Lipase	7.6–8.2	Olive oil	Aumaitre & Rérat (1968)
Cholesterol esterase	6.5–8.5	Cholesterol oleate	Momsen & Brockman (1977)
Phospholipase A_2	7.9–8.4	Egg yolk	de Haas et al. (1968)

Table 3.3 In vitro pH optima for hydrolases of porcine intestinal mucosa

Enzyme	Optimum pH	Substrate	Reference
Invertase	6.5	Sucrose	Dahlqvist (1960a)
Maltase	6.5–7.5	Maltose	Dahlqvist (1960b)
Lactase	6.0	Lactose	Dahlqvist (1961)
Trehalase	6.0	Trehalose	Dahlqvist (1960c)
Enterokinase	5.2–6.0	Trypsinogen	Neurath & Schwert (1950)
Dipeptidases	7.4–7.9	Various dipeptides	Josefsson & Lindberg (1965)

teases, the pH optimum is lower in the presence than in the absence of bile salt, and the figures at the lower end of the pH range are therefore the applicable values. The pH optimum for lipase is influenced by the presence of one of the products, free fatty acid. This inhibits the reaction, and its removal by ionization at a high pH raises the apparent optimum pH of the reaction. In vivo, fatty acids are removed by the mucosa, and the apparent optimum pH would depend on the rate of their removal. Table 3.3 gives the pH optima for those mucosal enzymes which are known to be located on the mucosal surface. It will be noted that lactase, trehalase and enterokinase which have the lowest pH optima are located mainly in the duodenum or proximal jejunum, where the pH is lowest.

It is unfortunate that so much enzyme assay work has been done without regard to the pH prevailing in the gut of the piglet. Lundh & Borgström (1962) commented on the difference between pH optima of pancreatic enzymes and the actual pH values which they recorded along the length of the human small intestine, the latter values being similar to those found by Walker (1959) and Smith & Jones (1963) in the small intestine of the sow-reared pig.

It seems that further comparisons, between digestive phsyiology of sow-reared piglets and piglets reared artificially would be likely to reveal interesting contrasts which, in turn, might well point the way towards the solution of many of the problems of artificial rearing.

In addition to the possible effect of pH within the gut on activity of digestive enzymes, pH may have other effects. For example, in vitro studies by Freeman (1969), intended to simulate conditions within the gut of the pig, showed that the distribution of fatty acids between micellar and oil phases was markedly affected by pH. Below a pH of 6.0, there was limited distribution of fatty acids in the micellar phase, and impairment of fatty acid absorption might be expected under such conditions in the gut. The partition of fatty acids in the micellar phase increased rapidly as pH increased above 6.2.

REFERENCES

Argenzio, R.A. & Southworth, M. (1974) Sites of organic acid production and absorption in the gastrointestinal tract of the pig. *Am. J. Physiol., 228,* 454–460.

Aumaitre, A. & Rérat, A. (1968) Proc. 2nd Wld Conf. Anim. Prod., College Park, Maryland, U.S.A., Session 7, Paper 83.

Braude, R., Fulford, R.J. & Low, A.G. (1976) Studies on digestion and absorption in the intestines of growing pigs. Measurements of the flow of digesta and pH. *Br. J. Nutr., 36,* 497–510.

Buraczewska, L., Buraczewski, S., Raczyński, G. & Zebrowska, T. (1973) Wplyw ogrzewania kazeiny w obecności glukozy na jej trawienie i wchlanianie w jelicie cienkim świń. *Roczn. Nauk Roln., 94–B–4,* 123–134.

Clemens, E.T., Stevens, C.E. & Southworth, M. (1975) Sites of organic acid production and pattern of digesta movement in the gastrointestinal tract of swine. *J. Nutr., 105,* 759–768.

Cranwell, P.D., Noakes, D.E. & Hill, K.J. (1968) Observations on the stomach contents of the sucking pig. *Proc. Nutr. Soc., 27,* 26A.

Cranwell, P.D., Noakes, D.E. & Hill, K.J. (1976) Gastric secretion and fermentation in the suckling pig. *Br. J. Nutr., 36,* 71–86.

Dahlqvist, A. (1960a) Characterisation of hog intestinal invertase as a glucosido-invertase. III. Specificity of purified invertase. *Acta chem. scand., 14,* 63–71.

Dahlqvist, A. (1960b) Characterisation of three different hog intestinal invertases. *Acta chem. scand., 14,* 1–8.

Dahlqvist, A. (1960c) Characterisation of hog intestinal trehalase. *Acta chem. scand., 14,* 9–16.

Dahlqvist, A. (1961) Pig intestinal β-glucosidase activities. I. Relation to β-galactosidase (lactase). *Biochim. biophys. Acta, 50,* 55–61.

Desnuelle, P. (1960) Chymotrypsin. *in* The enzymes. Edn. 2, vol 4, Eds Boyer, P.D., Lardy, H. & Myrbäck, K., pp. 93–118. Academic Press, New York.

Fischer, E.H. & Bernfeld, P. (1948) Sur les enzymes amylolytiques. VIII. L'α-amylase de pancréas de porc crystallisée. *Helv. chim. Acta, 31,* 1831–1839.

Folk, J.E. & Schirmer, E.W. (1963) The porcine pancreatic carboxypeptidase A system. I. Three forms of the active enzyme. *J. biol. Chem., 238,* 3884–3894.

Fossum, K. & Liven, E. (1974) The distribution of enzymes and bacteria in the small intestines of slaughter pigs. *Acta path. microbiol. scand. sect. B, 82,* 644–652.

Freeman, C.P. (1969) Properties of fatty acids in dispersions of emulsified lipid and bile salt and the significance of these properties in fat absorption in the pig and the sheep. *Br. J. Nutr., 23,* 249–263.

de Haas, G.H., Postema, N.M., Nieuwenhuizen, W. & van Deenen, L.L.M. (1968) Purification and properties of phospholipase A from porcine pancreas. *Biochim. biophys. Acta, 159,* 103–117.
Harpur, R.P. & Popkin, J.S. (1965) Osmolarity of blood and intestinal contents in the pig, guinea pig and Ascaris lumbricoides. *Can. J. Biochem., 43,* 1157–1169.
Hartman, P.A., Hays, V.W., Baker, R.O., Neagle, L.H. & Catron, D.V. (1961) Digestive enzyme development in the young pig. *J. Anim. Sci., 20,* 114–123.
Hecker, J.F. & Grovum, W.L. (1975) Rates of passage of digesta and water absorption along the large intestines of sheep, cows and pigs. *Aust. J. biol. Sci., 28,* 161–167.
Horszczaruk, F. (1971a) Badania nad trawienem u świń. II. Badania nad trawienem i wchlanianiem przy zastosowaniu metody ubojowej. *Biul. Inst. Genetiki Hodowli Zwierzat PAN, 21,* 117–136.
Horszczaruk, F. (1971b) Badania nad trawieniem u świń. III. Badania nad trawieniem i wchlanianiem prostych i mostkowych prezetok jelita cienkego. *Biul. Inst. Genetiki Hodowli Zwierzat PAN, 21,* 137–172.
Horszczaruk, F. (1971c) Badania nad trawieniem u świń. I. Badania z żastosowaniem przetok jelita ślepego i okreznici. *Biul. Inst. Genetiki Hodowli Zwierzat PAN, 21,* 101–115.
Josefsson, L. & Lindberg, T. (1965) Intestinal dipeptidases. I. Spectrophotometric determination and characterisation of dipeptidase activity in pig mucosa. *Biochim. biophys. Acta, 105,* 149–161.
Keys, J.E. & DeBarthe, J.V. (1974) Site and extent of carbohydrate, dry matter, energy and protein digestion and the rate of passage of grain diets in swine. *J. Anim. Sci., 39,* 57–67.
Kovács, F., Nagy, B. & Sinkovics, G. (1972) The gut bacterial flora of healthy early weaned piglets with special regard to factors influencing their composition. *Acta. vet. Hung., 22,* 327–338.
Kutas, F. & Szabó, J. (1974) A study of gastric acid production in pigs in relation to early weaning. *Acta vet. hung., 24,* 133–138.
Kutas, F., Szabó, J. & Ribiczeyné–Szabó, P. (1973) A korai elvalasztás emésztésélettani problémái malacokban. *Magyar Állatorvosok Lapja, 28,* 266–271.
Kvasnitskii, A.V. (1951) Voprosy fiziologii pishchevareniya u svinei. Sel'khozgiz, Moscow.
Laplace, J.P. (1974) Enregistrement du pH intragastrique chez le porc; variations liées à la nature, à l'importance et à l'intervalle de repas chez l'animal en finition. *Annls Zootech., 23,* 89–104.
Lawrence, T.L.J. (1970) Some effects of including differently processed barley in the diet of growing pigs. 2. In vivo gastric pH changes. *Animal Production, 12,* 151–163.
Lawrence, T.L.J. (1972) The effect of certain dietary factors on in vivo pH changes and pepsin activity in the stomach of the growing pig. *Br. vet. J., 128,* 402–411.
Lewis, U.J., Williams, D.E. & Brink, N.G. (1956) Pancreatic elastase: purification, properties and function. *J. biol. Chem., 222,* 705–720.
Lundh, G. & Borgström, B. (1962) Normal and abnormal digestive function. *in* The exocrine pancreas. pp 259–277. CIBA Foundation Symposium, Churchill, London.
Maner, J.H., Pond, W.G., Loosli, J.K. & Lowrey, R.S. (1962) Effect of isolated soybean protein and casein on the gastric pH and rate of passage of food residues in baby pigs. *J. Anim. Sci., 21,* 49–52.
Manners, M.J. (1970) Milk replacers for piglets. *J. Sci. Fd. Agric., 21,* 333–340.
Maxwell, C.V., Reimann, E.M., Hoekstra, W.G., Kowalczcyk, T., Benevenga, N.J. & Grummer, R.H. (1970) Effect of dietary particle size on lesion development and on the contents of various regions of swine stomach. *J. Anim. Sci., 30,* 911–922.
Mollgaard, H. (1946) On phytic acid, its importance in metabolism and its enzymic cleavage in bread supplemented with calcium.. *Biochem. J., 40,* 589–603.
Momsen, W.E. & Brockman, H.L., (1977) Purification and characterization of cholesterol esterase from porcine pancreas. *Biochim. biophys. Acta, 486,* 103–113.
Neurath, H. & Schwert, G.W. (1950) The mode of action of the crystalline pancreatic proteolytic enzymes. *Chem. Rev., 46,* 69–153.
Northrop, J.H., Kunitz, M. & Herriott, R.M. (1948) Crystalline enzymes. Edn 2. Columbia Univ. Press, New York.
Polivoda, D.I., Derevins'kii, V.V. & Kudrik, M.A. (1973) Shlunkove travleniya u porosyat pri rann'omu vidluchenni. *Svinovodstvo, Kiev, 26,* 68–73.

Schicketanz, W. & Richter, H. (1967) Orientierende Untersuchungen zum Verhalten der Wasserstoffionenkonzentration im Magen von Saugferkeln im Alter von 5 Wochen bis zum Zeitpunkt des Absetzens. *Mh. VetMed., 22,* 508–512.
Schulman, A. (1973) Effect of weaning on the pH changes of the contents of the stomach and duodenum. *Nord. VetMed., 25,* 220–225.
Sève, B & Laplace, J.P. (1975) Influence de la substitution des protéines du lait par des protéines de poisson sur quelques caracteristiques du contenu gastrique chez le porcelet sevré a dix jours. *Annls. Zootech., 24,* 43–57.
Slivitskii, M.G. (1975) Brodil'nye protsessy v zheludke svinei. *Vest. sel'khoz. Nauki., 7,* 75–80.
Smith, H.W. & Jones, J.E.T. (1963) Observations on the alimentary tract and its bacterial flora in healthy and diseased pigs. *J. Path. Bact., 86,* 387–412.
Stevens, J.A. & Kidder, D.E. (1972) The distribution of trehalase, sucrase, glucoamylase and lactase (β-galactosidase) along the small intestine of five pigs. *Br. J. Nutr., 28,* 129–137.
Walker, D.M. (1959) The development of the digestive system of the young animal. 1. Tissue weights, dry matter of tissues, total acidity and chloride content of stomach contents in the young pig. *J. agric. Sci., Camb., 52,* 352–356.
Wolff, E.C., Schirmer, E.W. & Folk, J.E. (1962) The kinetics of carboxypeptidase B activity. I. Kinetic parameters. *J. biol. Chem., 237,* 3094–3104.
Yamasaki, Y., Murakami, K., Irie, M. & Ukita, T. (1968) Purification and properties of porcine pancreatic ribonuclease. *J. Biochem. Tokyo, 63,* 25–32.
Zebrowska, T. (1973) Influence of dietary protein source on the rate of digestion in the small intestine of pigs. I. Amount and composition of digesta. *Roczn. Nauk Roln. 95–B–1,* 115–133.
Zebrowska, T. & Buraczewska, L. (1972) Wplyw poziomu bialka w diecie na ilość i sklad tresci jelita cienkego świń. *Zesz. Probl. postep. Nauk Roln., 126,* 115–121.

CHAPTER 4
GASTROINTESTINAL HORMONES

INTRODUCTION

Secretion and motility in the gastrointestinal tract are regulated by a combination of nervous and hormonal stimuli. The hormones involved are polypeptides produced by the mucosa of various parts of the tract and either act locally or are released into the blood, when they may act on the same or different parts of the digestive system.

The first demonstration of gastrointestinal hormone activity was by Bayliss & Starling (1902), who showed that a substance released into the blood from the duodenal mucosa of the dog stimulated pancreatic secretion. They called the substance "secretin", and showed (Bayliss & Starling, 1903) that the same type of action existed in a wide range of vertebrates including the pig. Starling (1905) coined the word "hormone" to describe the action of a substance released into the blood from one location in the body to act on another. Pig mucosa, because of its availability, became the usual source of secretin and other hormones of gastrointestinal origin. Another mucosal hormone, stimulating gastric secretion, was postulated by Edkins (1906) and called "gastrin", and one stimulating gall-bladder contraction, by Ivy & Oldberg (1928), which they called "cholecystokinin." Since then a large number of other hormones have been postulated as activators or inhibitors of various gastrointestinal functions, pig duodenal mucosa being the usual source of the active extracts.

In the absence, until recently, of satisfactory methods for separating peptides, none of these hormones was isolated pure for nearly 60 years after the initial discovery by Bayliss & Starling. Although the existence of secretin was accepted fairly quickly, the reality of gastrin and some of the other postulated hormones was questioned for a long time. This was not only because of the failure to isolate these substances, but also because the activities attributed to one hormone could frequently be elicited by extracts of another alleged hormone. When secretin, cholecystokinin (CCK) and gastrin were isolated pure, it was found that each of the hormones had several effects, and that some effects could be mediated by more than one hormone. For example, cholecystokinin stimulates contraction of the gall-bladder and increases enzyme secretion by the pancreas, two effects which were believed for a long time to be due to separate hormones. Conversely, some physiological effects can be mediated by more than one of the hormones. Gastric secretion of acid, for example, is inhibited both by secretin and by gastric inhibitory polypeptide.

The amino acid sequences of a number of gastrointestinal hormones have now been determined. In those in which the structure has been determined in more than one mammalian species, the hormone differs from one species to another by just one or two amino acids. Tracy & Gregory (1964) have shown that only the four amino acid C-terminal peptide sequence is essential for the main physiological actions of gastrin. Where the amino acid sequences of other gastro-intestinal hormones have been determined, usually on material from pig

UNIVERSITY
OF BRISTOL
VET. SCIENCE

mucosa, the different hormones have sometimes been found to have parts of the polypeptide chain in common, not only with one another, but also with other mammalian and even non-mammalian hormones which do not appear to have gastrointestinal origins or function such as caerulein, found in the skin of some amphibia. They can be divided into two groups, the "gastrin" group (Fig. 4.1), and the "secretin" group (Fig. 4.2). Dockray (1977) discussed the implications of these relationships.

The structural resemblances of the different hormones could explain their overlapping physiological activities, which are rather striking. Grossman (1973) listed the effects of gastrin, secretin, CCK and glucagon, a pancreatic hormone which raises the blood sugar level. He showed that if one looked at the motility of the stomach, intestines and gall-bladder, the exocrine secretions of the stomach, pancreas, liver and duodenum, and the blood supply to the intestines, each of the four hormones affected most of these parameters positively or negatively. The hormones which have not been examined so fully seem also to share some of these actions. Although some reported minor effects of individual hormones appear to have been due to contamination with another hormone, most of the substances, even when completely pure, have several physiological effects.

This overlapping of activities of different hormones means that when any new hormonal activity is reported, it is difficult to be sure that it is not due to an additional effect of one or more known hormones, or a breakdown product of one of them. Andersson (1973a) considered that before a substance can be added to the list of gastrointestinal hormones, it must be isolated pure, its effects studied and its release and action demonstrated under physiological conditions. Gastrin, secretin and CCK clearly fulfil these criteria, while several others in the long list of reported gastrointestinal hormones have been more or less sufficiently studied to come within this category.

The gastrointestinal hormones have recently been the subject of a textbook (Jorpes & Mutt, 1973) and several symposia (Demling, 1972; Andersson, 1973b; Chey & Brooks, 1974; Thomson 1975; Pickering, 1976) and review articles (Andersson, 1973a; Gregory, 1974; Johnson, 1974; Barrington & Dockray, 1976; Casthélaz, Griessen & Felber, 1977). In addition, there are review articles on specific aspects of the hormones. Grossman (1974) reviewed the less well established hormones, Makhlouf (1974) dealt with the interrelationship of the gastrointestinal hormones with the nervous system, Harvey (1975a) reviewed their effects on motility, and Dockray (1977) discussed the structural and interspecific relationships.

Makhlouf (1974) made it clear that the action of these hormones cannot be described fully without considering the autonomic nervous system, and that the hormones can be regarded in a sense as extensions of this system. At the site of hormone production the autonomic nervous system can provide the stimulus, or part of the stimulus for release of a hormone. At the site of action, the autonomic nervous system may be stimulated by the hormone or may potentiate the receptor organ so that it responds to the hormone. A further aspect of the controlling effect of the gastrointestinal hormones is shown by the discovery of a polypeptide, somatostatin which inhibits the release of some other polypeptide hormones.

In addition to the long-established effects on secretion, motility and blood flow, it is now clear that some at least of these hormones have a "trophic"

Pyr–Gln–Asp–Tyr–Thr–Gly–Trp–Met–Asp–Phe–NH_2
 |
SO_3H

–Asp–Tyr–Met–Gly–Trp–Met–Asp–Phe–NH_2
 |
SO_3H

–Glu–Ala–Tyr——— Gly–Trp–Met–Asp–Phe–NH_2
 |
SO_3H

Fig. 4.1 *The complete amino acid sequence of caerulein (top) and the C-terminal sequences of CCK (middle) and gastrin II (bottom) (Gregory, 1974)*

	1	2	3	4	5	6	7	8	9	10	11	12	13	14
VIP	His–	Ser–	Asp–	Ala–	Val–	Phe–	Thr–	Asp–	Asn–	Tyr–	Thr–	Arg–	Leu–	Arg–
Secretin	His–	Ser–	Asp–	Gly–	Thr–	Phe–	Thr–	Ser–	Glu–	Leu–	Ser–	Arg–	Leu–	Arg–
Glucagon	His–	Ser–	Gln–	Gly–	Thr–	Phe–	Thr–	Ser–	Asp–	Tyr–	Thr–	Lys–	Tyr–	Leu–
GIP	Tyr–	Ala–	Glu–	Gly–	Thr–	Phe–	Ile–	Ser–	Asp–	Tyr–	Ser–	Ile–	Ala–	Met–

	15	16	17	18	19	20	21	22	23	24	25	26	27	28	29
VIP	Lys–	Gln–	Met–	Ala–	Val–	Lys–	Lys–	Tyr–	Leu–	Asn–	Ser–	Ile–	Leu–	Asn–NH_2	
Secretin	Asp–	Ser–	Ala–	Arg–	Leu–	Gln–	Arg–	Leu–	Leu–	Gln–	Gly–	Leu–	Val–NH_2		
Glucagon	Asp–	Ser–	Arg–	Arg–	Ala–	Gln–	Asp–	Phe–	Val–	Gln–	Trp–	Leu–	Met–	Asp–	Thr
GIP	Asp–	Lys–	Ile–	Arg–	Gln–	Asp–	Phe–	Val–	Asn–	Trp–	Leu–	Ala–	Gln–	Gln–	

Fig. 4.2 *The amino acid sequences of the porcine hormones VIP, secretin, glucagon and GIP (only twenty-nine residues of GIP are shown) (Bodanszky et al., 1973)*

action, stimulating the growth of the receptor cells (Johnson, 1975, 1976). While this has been particularly well defined with regard to the effect of gastrin in stimulating growth of oxyntic cells in the gastric fundus (Johnson, Aures & Yuen, 1969; Johnson, 1975, 1976), there is evidence that other gastrointestinal hormones also have trophic effects on their respective target cells (Johnson, 1975). These trophic effects may be of importance in producing the secretory variations and adaptations in animals on different diets or different rearing regimes.

Table 4.1 gives a provisional list of gastrointestinal peptide hormones. We shall deal, as far as possible, with the presence and distribution of these hormones in the pig.

Table 4.1 *The peptide hormones of the gastrointestinal tract*

Hormone	Region of origin	Main effects	
Gastrin	Pyloric antrum	Gastric mucosa	Stimulates water and acid secretion
Secretin	Small intestine	Pancreas	Stimulates water and electrolyte secretion
Cholecystokinin-Pancreozymin	Small intestine	Stomach	Stimulates pepsinogen secretion
		Pancreas	Stimulates enzyme secretion
		Gall-bladder	Stimulates emptying
Gastric-inhibitory Peptide (G.I.P.)	Small intestine	Stomach	Inhibits secretion and motility
Vasoactive intestinal polypeptide (V.I.P.)	Small intestine	Stomach	Inhibits secretion and motility
Motilin	Small intestine	Small intestine	Increases gastric motility
Chymodenin	Small intestine	Pancreas	Stimulates chymotrypsinogen secretion
Bulbogastrone	Duodenal bulb	Stomach	Inhibits secretion and motility
Enteroglucagon	Small intestine	?	?
Somatostatin	Hypothalamus Pancreas Small intestine	?	Inhibits hormone secretion
Pancreatic polypeptide	Pancreas	Pancreas	Inhibits enzyme secretion
		Gall-bladder	Inhibits contraction

GASTRIN

A hormone from pig pyloric mucosa which stimulated hydrochloric acid secretion by the fundal region of the stomach was reported by Edkins (1906). After a long period in which doubts were expressed as to the reality of "gastrin" as this hormone was called, and a widely-accepted belief that its effects were really due to histamine, its existence was clearly demonstrated by Komarov (1938). In 1964 Gregory and Tracy isolated two peptide hormones with "gastrin" activity which they called gastrins I and II. The structure in Fig. 4.3 was established for gastrin II, while gastrin I lacked the sulphate ester group but was otherwise identical (Gregory, Hardy, Jones, Kenner & Sheppard, 1964).

$$\text{Glu–Gly–Pro–Try–Met–Glu–Glu–Glu–Glu–Glu–Ala–}\underset{}{\overset{SO_3H}{\overset{|}{\text{Tyr}}}}\text{–Gly–Try–Met–Asp–Phe–}NH_2$$

Fig.4.3 *Amino acid sequence of Gastrin II (Gregory et al., 1974)*

Tracy & Gregory (1964) showed with studies on synthetic peptides that only the C-terminal 4 amino acids in the correct sequence were necessary for gastrin activity. In addition to gastrins I and II, other larger and smaller molecules with gastrin activity have been found in human antral mucosa and in circulating blood. The larger of these, originally called "big" gastrins, break down on tryptic digestion to give gastrins I and II (Yalow & Berson, 1971), and have been shown to consist of two peptides, each of 34 amino acids, the last 17 of which are identical with gastrins I and II. Big gastrins have also been found in pig antral mucosa (Gregory & Tracy, 1972, 1973). The full amino acid sequences of these big gastrins are now known for the pig and man (Barrington & Dockray, 1976) and they are now called gastrin 34, or G34 and the 17-amino acid gastrins, G17. The G34 and the G17 appear to have the same actions but the G34 gastrins have a longer half-life in the blood (Walsh, 1975) and so will produce a more prolonged response after secretion. A still larger gastrin "big, big gastrin" has been reported (Yalow & Berson, 1971) and, from human material, traces of smaller "mini-gastrins" containing only the terminal 14 amino acids in gastrins I and II (Gregory & Tracy, 1972). In pig mucosa the sulphated form of the gastrins of each size (II) exceed the unsulphated (I) in the ratio of 2:1 (Gregory & Tracy, 1975).

Gastrin is present at high concentrations in the mucosa of the pyloric (antral) region of the stomach of the pig. In addition to its main effect of stimulating gastric acid and water secretion, gastrin has some of the effects of CCK, due presumably to its structural similarity. The effects of gastrin as given by Gregory (1974) are listed in Table 4.2.

The trophic effect of gastrin in stimulating growth of the acid-secreting oxyntic cells of the fundal mucosa of the stomach is now well established (Johnson, Aures & Yuen, 1969; Johnson, 1975, 1976). It has been suggested by Cranwell (1978) that this could be relevant to the earlier development of gastric acidity in piglets on some regimes than on others.

Table 4.2 *Actions of gastrin (Gregory, 1974)*

I. The following are produced by low doses

1. Stimulates secretion of water and electrolyte by stomach, pancreas, liver and Brunner's glands
2. Inhibits absorption of water and electrolytes from ileum.
3. Stimulates secretion of enzymes from stomach and pancreas.
4. Stimulates the lower oesophageal (cardiac) sphincter and stomach muscle.
5. Inhibits tone of the sphincter of Oddi.
6. Increases gastric mucosal blood-flow.
7. Releases histamine and induces histidine decarboxylase activity in gastric mucosa of rats.
8. Stimulates incorporation of amino acids into protein in gastric mucosa.

II. The following are produced by high doses

1. Stimulates the growth of gastric mucosa.
2. Stimulates insulin release.
3. Inhibits gastric secretion however excited (dogs only).
4. Stimulates smooth muscle of gut, gall-bladder, uterus.

Gastrin secretion is influenced by nervous stimuli, by hormones in the blood arising from other regions of the gastro-intestinal tract, and by conditions in the lumen of the antral (pyloric) part of the stomach. The main stimulant effects are: sensations associated with food, the presence in the antral lumen of particular amino acids (especially glycine, alanine and serine) or of volatile fatty acids (Makhlouf, 1974), and high antral pH. The principal inhibitor of gastrin release is low antral pH.

SECRETIN

Secretin activity was first demonstrated by Bayliss & Starling in the dog in 1902 and then shown to be present in other species including the pig (Bayliss & Starling, 1903). It was isolated pure by Jorpes & Mutt (1961) and found to be a peptide with 27 amino acids (Jorpes, Mutt, Magnusson & Steele, 1962), and the amino acid sequence in Fig. 4.2 was determined by Mutt, Jorpes & Magnusson (1970). In the pig it has been found predominantly in the mucosa of the proximal part of the duodenum, Mallanby (1926) reporting that the concentration of secretin in the pig small intestine decreased rapidly with increasing distance from the pylorus (Table 4.3).

Table 4.3 *Secretin activity extracted from pig small intestine mucosa at different distances from the pyloric sphincter (Mellanby, 1926)*

Distance (feet)	Units of activity
1	100
2	76
3	56
4	48
5	24
6	16
8	7
10	4

As described in Chapter 5, the most striking effect of secretin in the blood is to stimulate a large increase in water and bicarbonate secretion by the pancreas into the pancreatic ducts and by the liver into the bile ducts. The effects of secretin on the volume and composition of pig pancreatic juice and bile are described in Chapter 5. Secretin, in most mammals, has an inhibitory affect on gastrin-stimulated gastric secretion (Jorpes & Mutt, 1973) and an inhibitory effect on gastric emptying (Vagne & Andre, 1971). The effect in the pig has not been determined. The known effects as listed by Gregory (1974) are given in Table 4.4.

The strongest stimulus for the release of secretin into the blood is the presence of acid in the duodenal lumen (Meyer, Way & Grossman, 1970). Hong, Nakamura & Magee (1967) showed, in the pig, the effect of lowered pH in the duodenum producing a typical secretin response from the pancreas.

Table 4.4 *Actions of secretin (Gregory, 1974)*

I. The following are produced by low doses

1. Stimulates secretion of water and electrolytes from pancreas.
2. Stimulates secretion of water and electrolytes by liver (bile).
3. Inhibits lower oesophageal (cardiac) sphincter.
4. Inhibits gastric emptying.
5. Inhibits gastric acid secretion.
6. Stimulates gastric pepsin secretion.
7. Decreases duodenal motility.

II. The following are produced by high doses

1. Releases insulin.
2. Stimulates Brunner's glands (duodenum) in cats and dogs.
3. Stimulates renal excretion of water, Na and K in dogs.
4. Stimulates cardiac output and splanchnic blood flow in cats and dogs.
5. Stimulates lipolysis in fat cells.

CHOLECYSTOKININ-PANCREOZYMIN (CCK)

Ivy & Oldberg (1928) discovered in the duodenal mucosa a hormone which was liberated by food (especially fat), which caused the contraction and emptying of the gall bladder and named it cholecystokinin (CCK). Harper & Raper (1943) discovered a duodenal hormone which stimulated enzyme secretion by the pancreas and called this pancreozymin (PZ).

Mutt (1964) showed that the same substance was responsible for the effects attributed to both cholecystokinin and pancreozymin. The hormone then became known as cholecystokinin-pancreozymin (CCK-PZ), but it is now usually simply called cholecystokinin (CCK). Its full amino acid sequence has now been determined (Mutt & Jorpes, 1972, Fig. 4.4), and the C-terminal heptapeptide sequence found to be almost identical with that of gastrin II (Fig. 4.1).

While CCK is always obtained from pig duodenal mucosa, it is considered to be less exclusively confined to this region of the gut than is secretin (Jorpes & Mutt, 1973).

Lys—Ala—Pro—Ser—Gly—Arg—Val—Ser—Met—Ile—Lys—Asn—Leu—Gln—Ser—Leu—Asp—Pro—

Ser—His—Arg—Ile—Ser—Asp—Arg—Asp—Tyr—Met—Gly—Trp—Met—Asp—Phe—NH_2

(Tyr: (SO_3H))

Fig. 4.4 *The amino acid sequence of porcine cholecystokinin (Mutt & Jorpes, 1971)*

Most authors, studying man and the dog, have considered the most powerful stimulus to CCK release to be the presence in the duodenal lumen of fat hydrolysis products, unhydrolysed fat having no effect (e.g. Sircus, 1958). Other stimulants to CCK secretion are the presence in the duodenum of products of meat hydrolysis (Wang & Grossman, 1951) or dilute hydrochloric acid (Hong, Magee & Crewdson, 1956). Go, Hofmann & Summerskill, (1970) agreed that fatty acids in the duodenum were a much more powerful stimulus than glucose, but found that the essential amino acids phenylalanine, methionine and valine had an even greater effect on pancreatic enzyme secretion. The effect of intra-duodenal infusion of fatty acid in producing increased pancreatic enzyme secretion was demonstrated in the pig by Hong & Magee (1970) who also found that unhydrolysed fat had little or no effect (see pancreatic secretion, Chapter 5). The secretion of CCK has been reviewed by Harvey (1975b).

While the two main effects of CCK are those on gall bladder contraction and pancreatic enzyme secretion (described under secretion of the relevant organs), it also has other effects on the digestive tract. Perhaps as a consequence of its structural similarity to gastrin, it can have a gastrin-like action on the stomach or act as an inhibitor of gastrin action, depending on the precise effect being considered, the CCK concentration and the species. At low concentrations it stimulates gastric acid secretion but at high concentrations it is inhibitory (Murat & White, 1966; Magee & Nakamura, 1966). It has a stimulatory effect on pepsin secretion (Heitmann, Jungreis & Janowitz, 1967). The effects of CCK as summarized by Johnson (1974) are given in Table 4.5. Other effects are listed by Harvey (1975b) and include a marked relaxation of the sphincter of Oddi, some stimulation of bile secretion, a trophic effect on pancreatic acinar cells, and appetite suppression.

Table 4.5 *Effects of cholecystokinin (Johnson, 1974)*

Stomach		
	Acid secretion	Weak stimulation alone inhibition of gastrin
	Pepsin secretion	Stimulation
	Motility	Stimulation
Pancreas		
	Water and HCO_3 secretion	Stimulation
	Enzyme secretion	Strong stimulation
	Insulin release	Stimulation
Liver		
	Water and HCO_3 secretion	Stimulation
Intestine		
	Brunner's gland secretion	Stimulation
	Motility	Stimulation
	Absorption of H_2O, Na	Inhibition
Gall bladder		
	Contraction	Strong stimulation

OTHER GASTROINTESTINAL HORMONES

Apart from gastrin, secretin and CCK, a number of peptides with pharmacological action have been isolated from the mucosa of various parts of the digestive tract. Usually they affect one or more of the functions of secretion, motility and blood flow in some part of the tract. Several of these have been isolated pure and in some (VIP, GIP, motilin) the complete amino acid sequence of the porcine hormone has been determined. What is not certain about these peptides is which, if any, of their demonstrable actions are of real physiological consequence. It is also not clear what stimuli cause their release.

A further number of other hormones have been postulated on the basis of the activity of mucosal extracts which, after purification to various degrees, showed activity which could not easily be attributed to any of the better characterized hormones. The less well established gastrointestinal hormones were reviewed by Grossman (1974) and in the symposium by Thompson (1975). They will be reported on briefly here.

Gastric Inhibitory Polypeptide (GIP)

Tyr—Ala—Glu—Gly—Thr—Phe—Ile—Ser—Asp—
—Tyr—Ser—Ile—Ala—Met—Asp—Lys—Ile—Arg—Gln—
—Gln—Asp—Phe—Val—Asn—Trp—Leu—Leu—Ala—
—Gln—Gln—Lys—Gly—Lys—Lys—Ser—Asp—Trp—
—Lys—His—Asn—Ile—Thr—Gln

Fig. 4.5 The amino acid sequence of Gastro-inhibitory polypeptide (G.I.P.) (from Brown & Dryburgh, 1971)

This substance was isolated from pig duodenal and jejunal mucosa by Brown, Mutt & Pederson (1970) and its structure (Fig. 4.5) determined by Brown & Dryburgh (1971) shows it to be in the same family as secretin (Fig. 4.2). Its actions are discussed by Brown, Dryburgh, Moccia & Pederson (1975). These include:—

1) Inhibition of gastric acid secretion, whether this is induced by gastrin or histamine.
2) Inhibition of pepsin secretion.
3) Inhibition of gastric motility.
4) Stimulation of insulin release.
5) Stimulation of intestinal secretion.

The stimulation of insulin release by the pancreas is regarded as an important function of this hormone (Dupre, Ross, Watson & Brown, 1973).

Vasoactive Intestinal Polypeptide (VIP)

This polypeptide was isolated from pig small intestinal mucosa by Said & Mutt (1970) as the component of their mucosal extracts which caused peripheral vasodilation. It has an amino acid sequence similar to secretin (Said & Mutt, 1972, Fig. 4.2). In addition to the vasodilator effect which was responsible for the isolation and name, this peptide has a wide range of biological activity (Table 4.6, Said 1974, 1975; Makhlouf & Said, 1975).

Table 4.6 *The actions of vasoactive intestinal polypeptide (V.I.P.) (from Said, 1974)*

Vasodilation of the peripheral (systemic), splanchnic (especially hepatic), coronary and pulmonary vascular beds.

Stomach Inhibits histamine and gastrin stimulated acid and pepsin secretion.

Pancreas Stimulates electrolyte and water secretion. Augments response to secretin and CCK.

Liver Increases bile flow.

Gall bladder Relaxes bladder and counteracts CCK contractile effect.

Intestine Stimulates small intestine secretion.

Metabolism Stimulates lypolysis and glycogenolysis.

Phe—Val—Pro—Ile—Phe—Thr—Tyr—Gly—Glu—Leu—Gln—
Arg—Met—Gln—Glu—Lys—Glu—Arg—Asn—Lys—Gly—Gln

Fig. 4.6 *Amino acid sequence of Motilin (Schubert & Brown, 1974)*

Motilin

This substance was isolated by Brown, Cook & Dryburgh (1972) from commercial secretin from pig duodenal mucosa as a factor causing increased gastric motility. Its structure (Fig. 4.6) was determined by Schubert & Brown (1974). The isolation of this substance was initiated by the observation of a hormonal effect on gastric motility resulting from alkalization of the duodenal lumen (Brown, Johnson & Magee, 1966).

Motilin appears to be present at the highest concentration in pig jejunum, with lower levels in the duodenum, a trace in the ileum and none in the antrum or fundus (Brown & Dryburgh, 1975).

Chymodenin

Adelson & Rothman (1974) isolated from pig duodenal mucosa a polypeptide which, on injection into rats, caused a specific increase in chymotrypsin secretion by the pancreas. The amino acid composition has been provisionally determined (Adelson, 1975).

Enteroglucagon

A peptide fraction with biological activity resembling that of pancreatic glucagon in causing a depression of blood glucose, has been isolated from the small intestine mucosa of the dog, man and, later, from the ileum of the pig (Murphy, Elmore & Buchanan, 1971). Further study showed that two such peptides appeared to be present in pig duodenum. One of these seems to be very similar to, if not identical with pancreatic glucagon, while the other was a slightly smaller molecule which behaved in a similar way biologically and immunologically (Sasaki, Rubalcava, Srikant, Baetens, Orci & Unger, 1975).

Bulbogastrone

A factor inhibiting the gastric secretory response to food has been reported from extracts of the duodenal bulb of the dog, man, and more recently the pig (Andersson, Nilsson, Sjödin & Unvnäs, 1972), and called "bulbogastrone". The authors regarded it as a specific gastric inhibitory substance located predominantly in this short length of the tract.

Pancreatic Peptide Hormones

The two well-known pancreatic peptide hormones, insulin in the islet β-cells, and glucagon in the islet α-cells, are concerned with controls of metabolism and storage of glucose throughout the body. In addition, a gastrin-like peptide has been found in the Islet cells of pig pancreas (Pointner, Vatier, Dubraquet, Accary & Bonfils, 1972). This substance has the gastric secretagogue action of gastrin but does not have the effects of gastrin on bile secretion (Pointner & Flegel, 1973). It is not known under what circumstances it is released from the pancreas.

A polypeptide, now known as pancreatic polypeptide, has been isolated from the pancreas, initially of the bovine, and shown to be released into the blood after meals (Lin, Chance & Evans, 1973). It is a 36 amino acid polypeptide and has now been obtained from the pancreas of several mammals including the pig, and shown to have a number of effects on gastrointestinal secretion and motility (Lin & Chance, 1974). These include relaxation of the gall bladder, increase in tone of the bile duct, inhibition of motility of the pyloric sphincter, the ileum and ileocaecal valve, and stimulatory or inhibitory effects on pancreatic and gastric secretion depending on concentration (Lin & Chance, 1974).

Somatostatin

A peptide, isolated by Brazeau, Vale, Burgus, Ling, Butcher, Rivier & Guillemin (1973), from the hypothalamus was found to have an inhibitory effect on growth-hormone release, and was in consequence called growth-hormone release-inhibiting hormone (GH-RIH), or somatostatin. It has now been found to have an action inhibiting the release of a number of other hormones from various endocrine tissues, including a powerful effect on the gastric mucosa, inhibiting gastrin release into the plasma (Bloom, Mortimer, Thorner, Hall, Gomez-Pan, Roy, Russel, Coy, Kastin & Schally, 1974). In addition to its first detected source in the hypothalamus, it has been identified by immuno-cytochemical methods in cells of the pancreas, duodenum, jejunum and upper ileum of pig, man and dog (Polak, Pearse, Grimelius, Bloom and Arimura, 1975). It thus appears that in addition to a system for release of peptide hormones, a further control mechanism exists, operating by inhibiting hormone release but it is impossible at present to say which systems it affects under physiological conditions.

REFERENCES

Adelson, J.W. (1975) Chymodenin: an overview. *in* Gastrointestinal hormones. Ed. Thompson, J.C. pp. 563–574. University of Texas Press, Austin.

Adelson, J.W. & Rothman, S.S. (1974) Selective pancreatic enzyme secretion due to a new peptide called chymodenin. *Science, 183,* 1087–1089.

Andersson, S. (1973a) Frontiers of gastrointestinal hormone research. Almqvist & Wiksell, Stockholm.

Andersson, S. (1973b) Secretion of gastrointestinal hormones. *Ann. Rev. Physiol, 35,* 431–452.

Andersson, S., Nilsson, G., Sjödin, L. & Uvnäs, B. (1973) Mechanism of duodenal inhibition of gastric acid secretion. *in* Nobel Symposium XVI. Frontiers in Gastrointestinal Hormone Research. Ed. Andersson, S. pp. 223–238. Almqvist & Wiksell, Stockholm.

Barrington, E.J.W. & Dockray, G.J. (1976) Gastrointestinal hormones. *J. Endocr., 69,* 299–325.

Bayliss, W.M. & Starling, E.H. (1902) Mechanism of pancreatic secretion. *J. Physiol., Lond., 28,* 325–353.

Bayliss, W.M. & Starling, E.H. (1903) On the uniformity of the pancreatic mechanism in vertebrates. *J. Physiol. Lond., 29,* 174–180.

Bloom, S.R., Mortimer, C.H., Thorner, M.O., Hall, R., Gomez-Pan, A., Roy, V.M., Russel, R.C.G., Coy, D.H., Kastin, A.J. & Schally, A.V. (1974) Inhibition of gastrin and gastric acid secretion by growth-hormone release-inhibiting hormone. *Lancet, ii,* 1106–1109.

Bodanszky, M., Klausner, Y.S. & Said, S.I. (1973) Biological activities of synthetic peptides corresponding to fragments of and to the entire sequence of the vasoactive intestinal peptide. *Proc. Natl Acad. Sci. U.S.A., 70,* 382–384.

Brazeau, P., Vale, W., Burgus, R., Ling, N., Butcher, M., Rivier, J. & Guillemin, R. (1973) Hypothalamic polypeptide that inhibits the secretion of immunoreactive pituitary growth hormone. *Science, 179,* 77–79.

Brown, J.C., Cook, M.A. & Dryburgh, J.R. (1972) Motilin, a gastric motor activity-stimulating polypeptide: final purification, amino-acid composition and C-terminal residues. *Gastroenterology, 62,* 401–404.

Brown, J.C. & Dryburgh, J.R. (1971) A gastric inhibitory polypeptide. II The complete amino acid sequence. *Can. J. Biochem., 49,* 867–872.

Brown, J.C. & Dryburgh, J.R. (1975) Current status of motilin. *in* Gastrointestinal hormones. Ed. Thompson, J.C. pp. 549–554. University of Texas Press, Austin.

Brown, J.C., Dryburgh, J.R., Moccia, P. & Pederson, R.A. (1975) Current status of GIP. *in* Gastrointestinal hormones. Ed. Thompson, J.C. pp. 537–547. University of Texas Press, Austin.

Brown, J.C., Johnson, L.P. & Magee, D.F. (1966) Effect of duodenal alkalinization on gastric motility. *Gastroenterology, 50,* 333–339.

Brown, J.C., Mutt, V. & Pederson, R.A. (1970) Further purification of a polypeptide demonstrating enterogastrone activity. *J. Physiol., Lond., 209,* 57–64.

Casthélaz, M., Griessen, M. & Felber, J.-P. L'intestin: gland endocrine aux sécrétions multiples. *Schweiz. med. Wschr., 107,* 33–37.

Chey, W.Y. & Brooks, F.P. (1974) Endocrinology of the gut. Charles B. Slack inc., Thorofare, New Jersey.

Cranwell, P. D. (1978) Acid and pepsin secretion in young pigs reared solely by the sow or supplemented by solid food and weaned at 21 days. *Proc. Nutr. Soc., 36,* 142A.

Demling, L. (1972) Gastrointestinal hormones. Georg Thieme Verlag, Stuttgart.

Dockray, G.J. (1977) Molecular evolution of gut hormones: application of comparative studies on the regulation of digestion *Gastroenterology, 72,* 344-358.

Dupre, J., Ross, S.A., Warson, D., & Brown, J.C. (1973) Stimulation of insulin secretion by gastric inhibitory polypeptide in man. *J. clin. Endocr., 37,* 826–828.

Edkins, J.S. (1906) The chemical mechanism of gastric secretion. *J. Physiol., Lond., 34,* 133–144.

Go, G.L.W., Hofmann, A.F. & Summerskill, W.H.J. (1970) Pancreozymin assay in man based on pancreatic enzyme secretion; potency of specific amino acids and other digestive products. *J. clin. Invest., 49,* 1558–1564;

Gregory, H., Hardy, P.M., Jones, D.S., Kenner, G.W. & Sheppard, R.C. (1964) The antral hormone gastrin. *Nature, Lond., 204,* 931–933.

Gregory, R.A. (1974) The gastrointestinal hormones: a review of recent advances. *J. Physiol., Lond. 241,* 1–32.

Gregory, R.A. & Tracy, H.J. (1964) The constitution and properties of two gastrins extracted from hog antral mucosa. I The isolation of two gastrins from hog antral mucosa. *Gut, 5,* 103–107.
Gregory, R.A. & Tracy, H.J. (1972) Isolation of two "big gastrins" from Zollinger-Ellison tumour tissue. *Lancet ii,* 797–799.
Gregory, R.A. & Tracy, H.J. (1973) Big gastrin. *Mount Sinai J. Med., 40,* 359–364.
Gregory, R.A. & Tracy, H.J. (1975) Chemistry of the gastrins: some recent advances *in* Gastrointestinal hormones Ed. Thompson, J.C. pp. 13–24. University of Texas Press. Austin.
Grossman, M.I. (1973) Spectrum of biological action of gastrointestinal hormones. *in* Nobel Symposium XVI. Frontiers of gastrointestinal hormone research. Ed. Andersson, S., Almqvist & Wiksell, Stockholm.
Grossman, M.I. (1974) Candidate hormones of the gut. *Gastroenterology, 67,* 730–755.
Harper, A.A. & Raper, H.S. (1943) Pancreozymin, a stimulant of the secretion of pancreatic enzymes in extracts of the small intestine. *J. Physiol., Lond., 102,* 115–125.
Harvey, R.F. (1975a) Hormonal control of gastrointestinal motility. *Am. J. dig. Dis., 20,* 523–539.
Harvey, R.F. (1975b) The secretion and actions of cholecystokinin. *in* Modern trends in gastroenterology No. 5. Ed. Read, A.E. pp. 175–202. Butterworth, London.
Heitmann, P., Jungreis, A.M. & Janowitz, H.D. (1967) Effect of acetolazide and cholecys–tokinin on the secretion of pepsin from histamine-stimulated Heidenhain pouches. *Gastroenterology, 52,* 211–215.
Hong, S.S. & Magee, D.F. (1970) Pharmacological studies on the regulation of pancreatic secretion. *Annls. Surg., 172,* 41–48.
Hong, S.S., Magee, D.F. & Crewdson, F. (1956) The physiology of gall bladder evacuation. *Gastroenterology, 30,* 625–630.
Hong, S.S., Nakamura, M. & Magee, D.F. (1967) Relationship between duodenal pH and pancreatic secretion in dogs and pigs. *Ann. Surg., 166,* 778–782.
Ivy, A.C. & Oldberg, E.A. (1928) A hormone mechanism for gall bladder contraction and evacuation. *Am. J. Physiol., 86,* 599–613.
Johnson, L.R. (1974) Gastrointestinal Hormones *in* Gastrointestinal Physiology. Eds Jacobson, E.D. & Shanbour, L.L. pp. 1–43. Butterworth, London.
Johnson, L.R. (1975) Trophic action of gastrointestinal hormones. *in* Gastrointestinal hormones, Ed. Thompson, J.C. pp. 215–230. University of Texas Press, Austin.
Johnson, L.R. (1976) The trophic action of gastrointestinal hormones. *Gastroenterology, 70,* 278–286.
Johnson, L.R., Aures, D. & Yuen, L. (1969) Pentagastrin-induced stimulation of protein synthesis in the gastrointestinal tract. *Am. J. Physiol., 217,* 251–254.
Jorpes, J.E. & Mutt, V. (1961) On the biological activity and amino acid composition of secretin. *Acta chem. scand., 15,* 1790–1791.
Jorpes, J.E. & Mutt, V. (1973) Secretin, cholecystokinin, pancreozymin and gastrin. *Handbuch der experimentellen Pharmakologie, NS 34,* Hefftner Heubner.
Jorpes, J.E., Mutt, V., Magnusson, S. & Steele, B. (1962) Amino acid composition and N-terminal amino acid sequence of porcine secretin. *Biochem. Biophys. Res. Commun., 9,* 275–279.
Komarov, S.A. (1938) Gastrin. *Proc. Soc. exp. Biol., 38,* 514–516.
Lin, T-M. & Chance, R.E. (1974) Gastrointestinal actions of a new bovine pancreatic peptide (BPP). *In* Endocrinology of the gut. Eds Chey, W.Y. & Brooks, F.P. pp. 143–145. Charles B. Slack inc., Thorofare, New Jersey.
Lin, T-M., Chance, R. & Evans, D. (1973) Stimulatory and inhibitory actions of a bovine pancreatic peptide (BPP) on gastric and pancreatic secretions in dogs. *Gastroenterology, 64,* 865.
Magee, D.F. & Nakamura, M. (1966) Action of pancreozymin preparations on gastric secretion. *Nature, 212,* 1487–1488.
Makhlouf, G.M. (1974) The neuroendocrine design of the gut. *Gastroenterology, 67,* 159–184.
Makhlouf, G.M. & Said, S.I. (1975) The effect of vasoactive intestinal peptide (VIP) on digestive and hormonal function. *in* Gastrointestinal hormones. Ed. Thompson, J.C. pp. 599–610. Texas University Press, Austin.
Mellanby, J.M. (1926) Secretion of pancreatic juice. *J. Physiol., Lond., 61,*419–435.
Meyer, J.H., Way, L.W. & Grossman, M.I. (1970) Pancreatic bicarbonate response to various acids in the duodenum of the dog. *Am. J. Physiol., 219,* 964–970.

Murat, J.E. & White T.T. (1966) Stimulation of gastric secretion by commercial cholecystokinin extracts. *Proc. Soc. exp. Biol. N.Y., 123,* 593–594.
Murphy, R.F., Elmore, D.T. & Buchanan, K.D. (1971) Isolation of a glucagon-like immunoreactivity of gut by affinity chromatography. *Biochem. J., 125,* 61P–62P.
Mutt, V. (1964) Behaviour of secretin, cholecystokinin and pancreozymin to oxidation with hydrogen peroxide. *Acta chem. scand., 18,* 2185–2186.
Mutt, V. & Jorpes, J.E. (1972) Hormonal polypeptides of the upper intestine. *Biochem. J., 125,* 57P–58P.
Mutt, V., Jorpes, J.E. & Magnusson, S. (1970) Structure of porcine secretin. The amino acid sequence. *Eur. J. Biochem. 15,* 513–519.
Pickering, B:T. (1976) The gut as an endocrine organ. *J. Endocrinol., 70,* 1P–15P.
Pointner, H. & Flegel, U. (1973) Effect of a gastro-secretory substance from hog pancreas (GSP) and gastrin on bile secretion in the rat. *Pflügers Archiv., 345,* 73–79.
Pointner, H., Vatier, J., Dubrasquet, M., Accary, J.P. & Bonfils, S. (1972) Isolation of a gastro-secretagogue substance from hog pancreas. *Clin. Chim. Acta., 42,* 321–326.
Polak, J.M., Pearse, A.G.E., Grimelius, L., Bloom, S.R. & Arimura, A. (1975) Growth-hormone release-inhibiting hormone in gastrointestinal and pancreatic D cells. *Lancet, i.,* 1220–1222.
Said, S.I. (1974) Candidate hormones of the gut. 5. Vasoactive intestinal peptide (V.I.P.) *Gastroenterology, 67,* 735–737.
Said, S.I. (1975) Vasoactive intestinal polypeptide: current status. *in* Gastrointestinal hormones. Ed. Thompson, J.C. pp. 591–597. Texas University Press, Austin.
Said, S.I. & Mutt, V. (1970) Polypeptide with broad biological activity; isolation from small intestine. *Science, 169,* 1217–1218.
Said, S.I. & Mutt, V. (1972) Isolation from porcine-intestinal wall of a vasoactive octacosapeptide related to secretin and to glucagon. *Eur. J. Biochem., 28,* 199–204.
Sasaki, H., Rubalcava, B., Srikant, C.B., Baetens, D., Orci, L. & Unger, R.H. (1975) Gut glucagonoid (GLI) and gut glucagon. *in* Gastrointestinal hormones. Ed. Thompson, J.C. pp. 519–528. Texas University Press, Austin.
Schubert, H. & Brown, J.C. (1974) Correction to the amino acid sequence of porcine motilin. *Can. J.Biochem., 52,* 7–8.
Sircus, W. (1958) Studies on the mechanism of the duodenum inhibiting gastric secretion. *Quart. J. exp. Physiol., 43,* 114–133.
Starling, E.H. (1905) On the chemical correlation of functions of the body. *Lancet, ii,* 339–341.
Thompson, J.C. (1975) Gastrointestinal hormones. University of Texas Press, Austin.
Tracy, H.J. & Gregory, R.A. (1964) Physiological properties of a series of synthetic peptides structurally related to gastrin I. *Nature, Lond., 204,* 935–938.
Vagne, M. & Andre, C. (1971) The effect of secretin on gastric emptying in man. *Gastroenterology, 60,* 421–424.
Walsh, J.H. (1975) Biological activity and disappearance rates of big, little and mini-gastrins in dog and man. *in* Gastrointestinal hormones. Ed. Thompson, J.C. pp. 75–83. University of Texas Press, Austin.
Wang, C.C. & Grossman, M.I. (1951) Physiological determination of release of secretin and pancreozymin from intestine of dogs with transplanted pancreas. *Am. J. Physiol., 164,* 527–544.
Yalow, R.S. & Berson, S.A. (1971) Further studies on the nature of immunoreactive gastrin in human plasma. *Gastroenterology, 60,* 203–214.

CHAPTER 5
SECRETION

INTRODUCTION

There is secretion into the lumen of the gastrointestional tract by three organs associated with the tract, namely the salivary glands, the pancreas and the liver, these secretions entering the tract by the salivary, pancreatic and bile ducts respectively. The wall of the tract is highly glandular in the fundal and pyloric regions of the stomach and in the duodenum, and considerable secretion into the lumen occurs from the mucosa in these regions. Water and electrolyte transport can take place in both directions through the mucosa of the small and large intestines, net secretion into the lumen normally occurring in the proximal small intestine. In addition, there are mucus-secreting cells along the entire length of the tract. Fig. 5.1 illustrates these secretions diagrammatically.

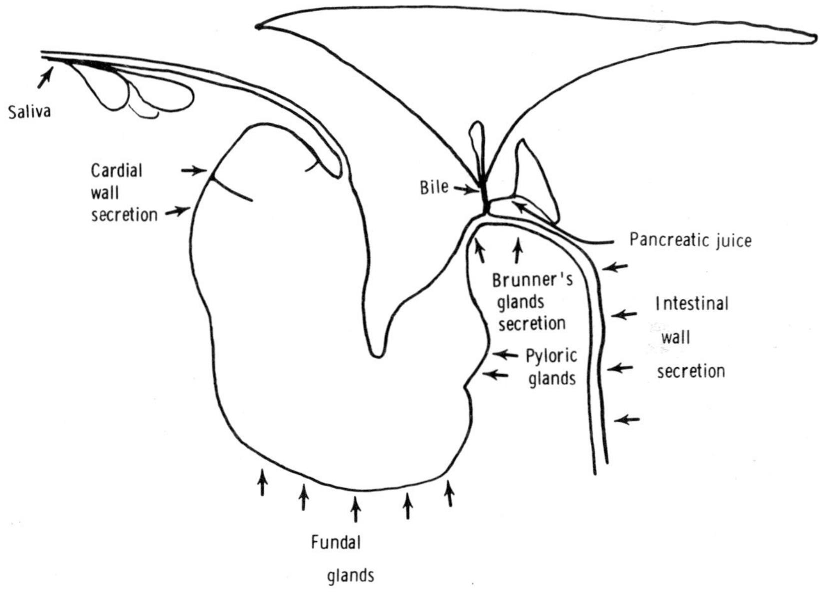

Fig. 5.1 Secretions into the gastrointestinal tract

The salivary glands, the pancreas and the glandular part of the stomach secrete digestive enzymes which break the food down to smaller molecules, and the liver secretes bile salts with surface active properties which assist the dispersion of fat and especially the micellar dispersion of the products of fat hydrolysis, which is essential to their absorption.

The saliva, the pancreatic juice, the bile and the Brunner's glands and other upper intestinal secretions contain bicarbonate, while the gastric secretion is strongly acid. Thus, each of these secretions helps to control the pH of the relevant portion of the gut lumen at a level appropriate to the digestive processes which occur in that region.

The secretions of the various glands are regulated by a combination of vagal, nervous and hormonal controls, which simultaneously regulate motility throughout the tract. Each gland has nerves which can supply stimulatory, and ones which can supply inhibitory impulses. Similarly, as described in Chapter 4, it can receive either stimulatory or inhibitory impulses from the hormones either released locally or arriving in its blood supply. The nervous impulses can be psychic in origin (the smell, sight, taste or thought of food), or can be feedback from receptors in another part of the tract. The hormones mainly originate in the mucosa and, on release, act on the same or another region of the tract. Most of the hormones have a combination of effects, stimulating one function and inhibiting another. The same is also, probably, the case with the nervous stimuli.

The overall effect of these control mechanisms is to provide secretions in quantities related to the requirements of the relevant portions of the tract at any particular time.

SALIVA

Composition

Saliva in mammals has the functions of moistening food, lubricating the oesophagus, and starting the digestion of starch. For these purposes the important constituents are water, mucus and α-amylase. Most of the studies on the composition and secretion of saliva have been made on man and the dog, but Kudryavtsev (1935) and Kvasnitskii (1951) have shown that pig salivary secretion is essentially similar. In studies on two pigs with cannulated parotid salivary glands, Kudryavtsev (1935) showed that the volume of salivary secretion varied in response to food or other stimulants in the mouth. In more extensive studies, Kvasnitskii (1951) confirmed this, and showed that as the rate of secretion increased, so the specific gravity, dry matter and amylase content fell. He found that the specific gravity varied from 0.999 to 1.0086, dry matter from 0.34% to 4.92% and amylase activity from 2 to 500 Wohlgemuth units. When a salivary gland had not been secreting for a time, the duct contained a jelly-like material with a relatively high dry-matter and high amylase content. Stimulation of secretion caused this to be flushed out by a thin watery saliva.

Arkhipovets (1956) fitted piglets with fistulas of the parotid salivary gland and studied the secretion during feeding on four different cereal diets over the age range 47–115 days. He recorded irregular variations in concentration from day to day but a slight tendency for specific gravity, dry matter, nitrogen content and amylase to increase with age. Under these conditions of stimulated secretion, specific gravity was in the range 1.0013–1.0058, dry matter 0.67–1.55%, nitrogen 0.40–1.05 mg/ml and amylase 12–64 Wohlgemuth units.

Volume

Kvasnitskii (1951) found that salivary secretion in the pig varied in speed of onset, duration of secretion and volume secreted, according to the stimulant. Five per cent HC1 produced secretion in about 5 seconds, the pig's usual cereal food in 6–8 seconds, and 5% butyric or formic acid in 14–27 seconds. Secretion was completely suppressed by washing out the mouth with water, and very little occurred with moist foods such as boiled potatoes. Secretion was intense and prolonged after feeding dry meal, the dryness of the food being apparently the most important stimulus. Dilute acids (hydrochloric, lactic, formic, acetic) all produced marked stimulation of secretion, but none produced a response comparable in volume or duration with that from a dry cereal mixture.

The left and right submaxillary glands seemed to respond independently and only to a stimulus on their own side of the mouth. Thus flushing the left side of the mouth with water and the right side with dilute acid produced salivary secretion from the right submaxillary gland and not from the left.

During a prolonged feed, salivary secretion reached the maximum level almost immediately, and continued at around this level for about an hour and then diminished slightly over the next hour, the composition of the saliva remaining unchanged. The total volume secreted by a sow under such conditions was estimated at 1.5 litres per gland per feed, giving a rough estimate of 15–18 litres for the animal per day.

Arkhipovets (1956), studying the flow of parotid saliva during feeding on four different cereal diets over the age range of 47–115 days, found that the flow rate increased with age roughly proportionately to increased food intake.

Enzyme activity

α-Amylase

α-Amylase is present in high concentration in pig saliva, and has the same enzymic properties as human salivary amylase, which breaks down starch to maltose, maltotriose and dextrins as described in Chapter 6.

Polivoda & Derevins'kii (1974) found that although the weight of the salivary glands of piglets as a proportion of body weight fell sharply from birth to 3 weeks of age, the total amylase content of the glands rose in proportion to body weight in the same period. From 3 weeks to 60 days of age, the weight of the glands dropped only very slightly as a proportion of the body weight but the amylase content in relation to body weight fell considerably in sow-reared piglets over this period but rather less in piglets weaned at 10 days.

Other enzymes

While traces of other enzymes have been found in saliva, they may be just leakage of tissue constituents and are probably of no digestive significance.

GASTRIC SECRETION

There are two entirely different types of secretion in the stomach of the pig, one, an alkaline secretion from the cardial region is very limited in volume while the other, an acid secretion from the fundal and pyloric regions, is produced in large quantities after ingestion of food.

Secretion from the cardial mucosa

In the first or cardial part of the pig's stomach, including the blind sac, there is a surface epithelium quite different from the glandular fundus and pylorus. The cardial type of epithelium covers 30–40% of the surface area of the stomach of a mature pig (Höller, 1970a). By surgically isolating and exteriorizing the blind sac in sixteen 30–40 kg pigs, Höller (1970a, b) studied the secretion of this region of the stomach. He found that the cation composition was similar to that of plasma, but more variable, and the anion composition differed in that the bicarbonate was about four times and the chloride about half that of a plasma ultrafiltrate (Höller, 1970a) and the secretion had a pH of 7.76±0.09 so that it bears more resemblance to a ruminal than to a gastric fundal secretion. The volume secreted over 24 hours in 30–40 kg pigs ranged from 50–340 ml, with a mean of 125 ml. Most of this was secreted during the night, a fact which could have been related to the feeding pattern, as feeding suppressed secretion by this region for some hours (Höller, 1970a). Treatments stimulating gastric acid secretion, such as injection of histamine or a gastrin analogue also caused suppression of the secretion of the blind sac epithelium (Höller, 1970b).

Secretion of the fundal and pyloric regions

Secretion by the fundal and pyloric regions of the pig's stomach is similar to that in other mammals, both in composition and in the control mechanisms, but in the pig the greater part of the acid secretion occurs in the fundus.

Secretion from the fundus of the stomach differs from other secretions in the body in that it is very acid and contains proteolytic enzymes which have their maximum activity at a very acid pH. In addition, the gastric mucosa secretes small quantities of a few other enzymes.

Gastric enzymes

PEPSIN

Pepsins are secreted as inactive precursors (pepsinogens) whose activation is initiated by the acid conditions in the stomach. They have been reviewed by Bovey & Yanari (1960) and Taylor (1962).

The main pepsin from pig gastric juice was purified by Northrop in 1930, and the corresponding pepsinogen by Heriott in 1938. Since then, a number of similar minor components of pig gastric juice have been isolated (Taylor, 1962), and the main component is now frequently referred to as pepsinogen A, the others being called pepsinogens B, C and D (Protein digestion Table 7.1). The amino acid sequence of pepsin has now been completely determined and the existence of 3 minor variants demonstrated (Sepulveda, Marciniczyn, Liu & Tang, 1975).

OTHER ENZYMES

Rennin

The milk-clotting protease rennin is produced in the gastric mucosa of the young piglet, the amount decreasing with age (Antonio, 1940).

Cathepsin

Some proteolytic enzymic activity with a pH optimum in the range 6.8–7.2 is present in the gastric juice of the pig, as it is in other species (Taylor, 1959). Its digestive function, if any, is not known.

Chitinase

The gastric mucosa of the pig was found by Jeuniaux (1961) to secrete chitinase, which will hydrolyse chitin, the structural amino-polysaccharide of the exoskeleton of insects and arthropods. He reported that while most insectivorous mammals produce this enzyme in either the gastric mucosa or the pancreas, the pig produces it in both.

Control of gastric secretion

The control of gastric secretion has been studied extensively in the dog, the cat, in man and in laboratory rodents, the same principal features always being present. The relatively limited studies on the pig indicate that the same mechanisms are present as in other mammals. These are described in text books on human and mammalian physiology and in more detail by James (1957) and Gregory (1962).

The most powerful stimulus to fluid and acid secretion is histamine, the presence of even very low levels in the blood causing copious secretion of very acid gastric juice, mainly from the fundal region of the stomach. Histamine is now regarded as the immediate stimulant to secretion by the acid-secreting oxyntic cells, other stimulants acting by causing production and release of histamine in the region of the fundal mucosa. The stimuli which are believed to act through histamine are vagus nerve stimulation and the hormone gastrin, which originates from the mucosa of the pyloric region of the stomach (Chapter 4). The nervous and hormonal stimuli act jointly, in that gastrin only has a limited effect in the absence of stimulation by the gastric vagus nerve.

The first secretory stimulus on feeding is the "cephalic" phase acting through the vagus nerves due to the thought of food in response to sights, sounds, smells or tastes associated with feeding. When food reaches the stomach, there is direct chemical action due to various constituents, but particularly peptides, on the mucosa of the pyloric region of the stomach, causing release of gastrin into the blood. This strongly stimulates the oxyntic cells, which are mainly in the fundal region, to secrete fluid and acid. This is the so-called "gastric" phase of secretion. Finally, food in the small intestine stimulates the release into the blood of other hormones. Some of these, especially secretin, inhibit gastric acid secretion, while both secretin and cholecystokinin (CCK) stimulate pepsin secretion. This is the so-called "intestinal" phase of gastric secretion.

In addition to the hormonal stimuli from the intestine, there are vagal nervous connections between the small intestine and the stomach, and if these are severed, the response of the stomach to the intestinal hormones is reduced.

Pepsin is secreted in response to the presence of some food substances, especially peptides, in the stomach, to vagal nerve stimulation, and to some of the intestinal peptide hormones, especially gastrin, CCK and secretin.

Although the studies on gastric secretion in the pig are relatively few compared with those on the dog and man, they do cover the basic secretory mechanisms.

It has been shown in fistularized pigs by Huang & Liu (1964), Noakes, Cranwell

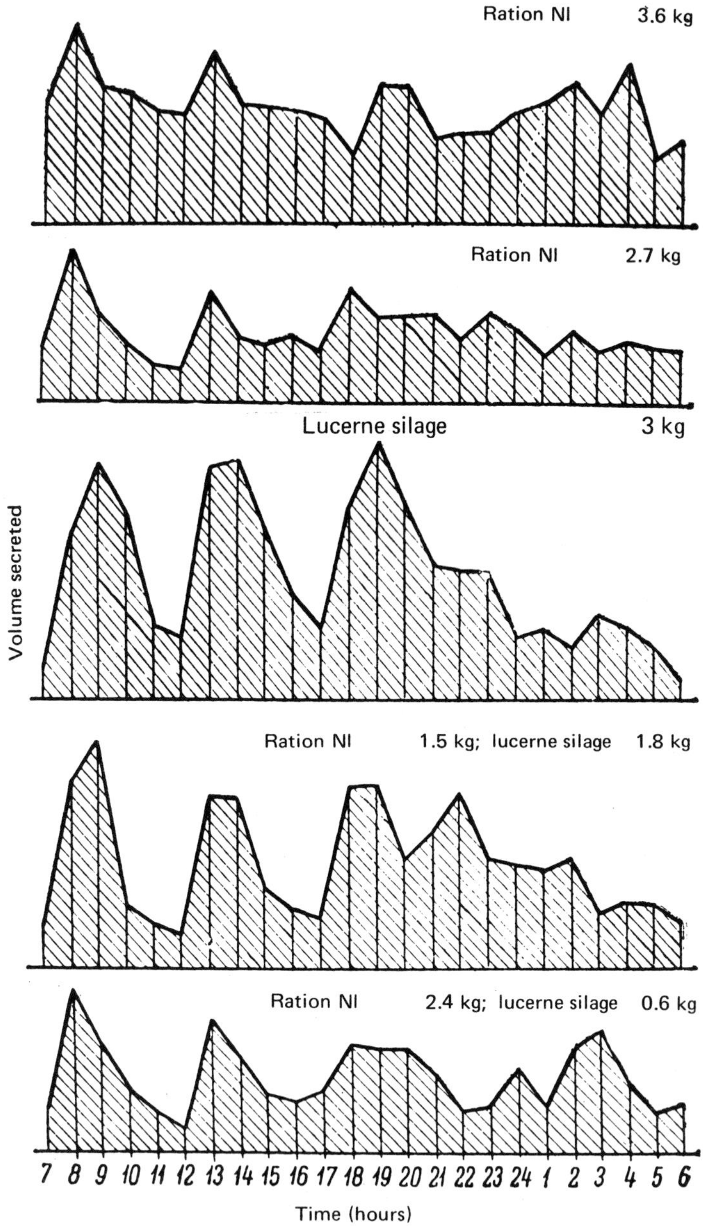

Fig. 5.3 24 hour secretion of gastric juice from the isolated stomach of a pig fed a cereal ration (ration NI) and silage. The pigs were fed at 7, 12 and 17 hours. (Kvasnitskii, 1951).

& Hill (1968), Merritt & Brooks (1970), Noakes (1972) and Moeller, Sewing, Geisbe & Schmolke (1974), that histamine injection produces profuse fluid and acid secretion, as it does in other species. This response has also been shown in an isolated perfused stomach by Kowalewski, Secord & Kolodej (1974). Kvasnitskii (1951) and Cheredkova (1959) demonstrated that sham feeding in a pig with a re-entrant oesophageal fistula stimulated gastric secretion of fluid, acid and pepsin. Cheredkova (1959) showed that this reaction was abolished in an isolated gastric pouch if the nervous connection was severed, though if food was placed in the stomach, the isolated pouch would respond by secreting fluid, acid and pepsin. Thus, the gastric fundus of the pig responds to a hormonal stimulus in the same way as that of the dog.

Volume of gastric secretion

Systematic studies using surgically isolated pouches and also analysis of stomach contents have been reported by Kvasnitskii (1951). He showed that while secretion continued throughout the 24 hours, it was much greater during the day than the night, being greatest in the 2–3 hours following a feed (Fig. 5.2). Changing the feeding regime so that the animals were fed during the night reversed the pattern (Table 5.1). The nature and quantity of the food also influenced the volume of juice secreted (Table 5.2; Fig. 5.3). Kvasnitskii demonstrated a marked increase in volume when silage was included in the diet, and Bakeeva (1962) noted a large increase with meat and bone meal.

Noakes (1972), in experiments on pigs of 8–14 weeks of age with either gastric fistulas or Heidenhain pouches, confirmed these findings of Kvasnitskii on gastric secretion.

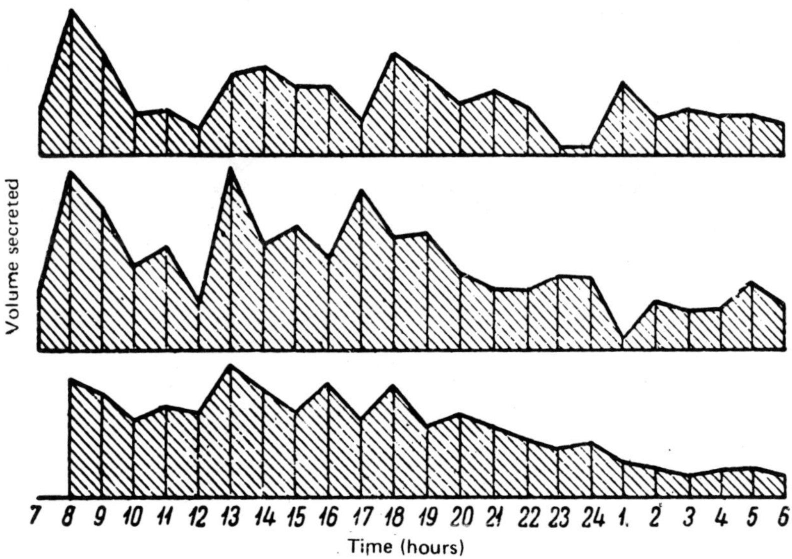

Fig. 5.2 Secretion of gastric juice from an isolated stomach over the course of 24 hours by 3 pigs. The pigs were fed at 7.00, 12.00 and 17.00 hours (Kvasnitskii, 1951).

Table 5.1 *Volume of gastric secretion (in cm³) from gastric pouches on 2 adult pigs when fed during the day, and during the night (Kvasnitskii, 1951).*

	Fed during the day			Fed during the night		
		Secretion			Secretion	
	Observation no.	day	night	Observation no.	day	night
Pig A	1	84.4	64.1	4	66.4	96.6
	2	130.4	123.2	5	86.5	106.0
	3	272.0	170.0	6	109.0	110.0
Pig B	1	198.4	105.4	3	129.0	147.2
	2	259.8	114.0	4	116.0	239.5

Table 5.2 *Volume of gastric secretion from an isolated stomach with feeds of different duration (Kvasnitskii, 1951).*

	Duration of feeding (mins)	
Type of ration	10	30
	Volume of juice (cm³)	
Barley	249	272
Green lucerne	195	267
Cow's milk	135	234

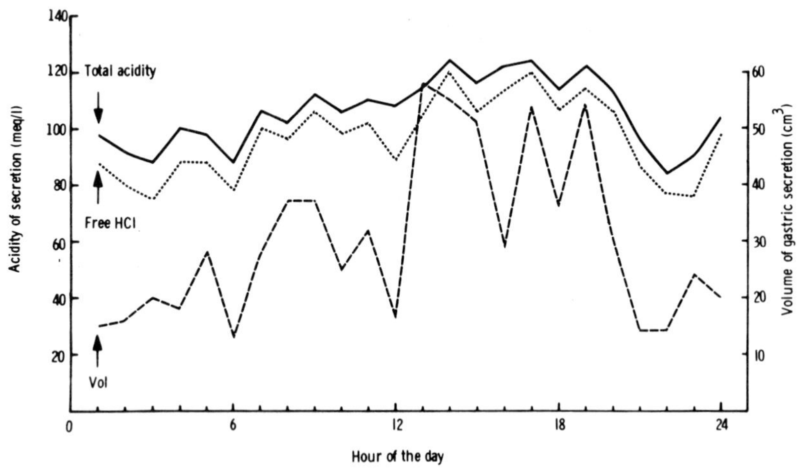

Fig. 5.4 *Relationship between acidity and volume of secretion from an isolated gastric pouch in a pig (from Kvasnitskii, 1951).*

Acidity and peptic activity

EFFECT OF RATE OF SECRETION

Kvasnitskii (1951) showed that the acidity of gastric juice normally increased with increase in the rate of secretion (Fig. 5.4). He found that the peptic activity, on the other hand, did not show any close correlation with volume or acidity, but, on the whole, was lower during periods of increased secretion. This effect on acid secretion was confirmed by Noakes (1972) who obtained free acidities of 100–150 meq/1, usually above 130 meq/1, for gastric juice from isolated pouches during feeding in 8–14 week old pigs. Noakes found the peptic activity to increase initially after stimulation and then to fall, probably due to a "washing out" effect of the continued high secretion volume.

Effect of diet on gastric secretion

Kvasnitskii (1951) studied this in pigs with isolated gastric pouches, where the quantity and composition of the secreted fluid was measured, and also with pigs with simple gastric fistulas, measuring the acidity of the contents. With both techniques, the addition of silage or of yeast to the diet increased the secretion of acid considerably.

Gridin (1956) confirmed the different gastric response to different feeds, and showed the way a bulky ration produced as high an initial response to feeding as a barley ration, but that the secretory rate fell off very rapidly after the bulky ration but remained high after the barley feeds. Bakeeva (1962), using pigs with isolated pouches, showed that as protein-containing constituents were added to a maize-based ration, so the output of fluid, free acid and pepsin increased. Bakeeva & Utekhin (1958) reported that the gastric secretory response to a feed was also depend on the diet to which the animal had been previously accustomed.

Development of secretion

FLUID AND ACID

The most extensive studies of the development of gastric secretion in the pig were those of Kvasnitskii & Bakeeva (1941), reported by Kvasnitskii (1951). In this work, the secretion of acid was found to be low for the first few weeks of life, and free acid (that is, a pH below 4.5) could not be detected until the pigs were over 3 weeks of age. From 4 weeks onwards, the acidity increased rapidly, and by 2–2½ months of age it approached that of adult pigs.

Walker (1959), in slaughter studies, also found free acid to be absent in the first 3 weeks, but Kudryavtsev, Kudryashov & Troyanovskii (1935), Cranwell, Noakes & Hill (1968; 1976) and Cranwell & Titchen (1974) all found free acid present almost from birth. As described in Chapter 6, Cranwell, Noakes & Hill (1968; 1976) showed that piglets suckled in a clean environment started to secrete gastric hydrochloric acid within a day or two of birth, but that piglets suckled in a conventional farm environment usually acquired a gastric population of lactobacilli which produced lactic acid and so lowered the pH to a level at which hydrochloric acid secretion was not stimulated until the piglets were about 3 weeks of age. This suppression of secretion probably accounts for the absence of free hydrochloric acid in pigs under 3 weeks of age reported by Kvasnitskii and Walker.

Kvasnitskii (1951) reported the absence of a reflex phase in gastric secretion in the suckling pig, and claimed that this response gradually developed after weaning, but Noakes (1972) showed that it existed in the first few weeks of life. Kutas & Szabó (1974) found that hydrochloric acid was secreted in the stomachs of only a small proportion of suckled piglets of 4 weeks of age in response to histamine injection, but that the great majority of early-weaned piglets of the same age secreted hydrochloric acid under the same conditions.

Cranwell (1978) suggested that the trophic effect of gastrin in stimulating the growth of acid-secreting cells could contribute to the further development of acid secretion in those piglets in which it had become established. Such an effect would help to explain both the results of Cranwell, Noakes & Hill and those of Kutas & Szabó.

Kvasnitskii (1951) found that the volume of gastric juice secreted per day increased with age, rapidly for the first few weeks and then more slowly, an observation which was confirmed by Noakes (1972). Kvasnitskii also found that, in the suckling pig, secretion was greater during the night than during the day, in contrast to the situation in the adult. The weaned pig was intermediate (Fig. 5.5).

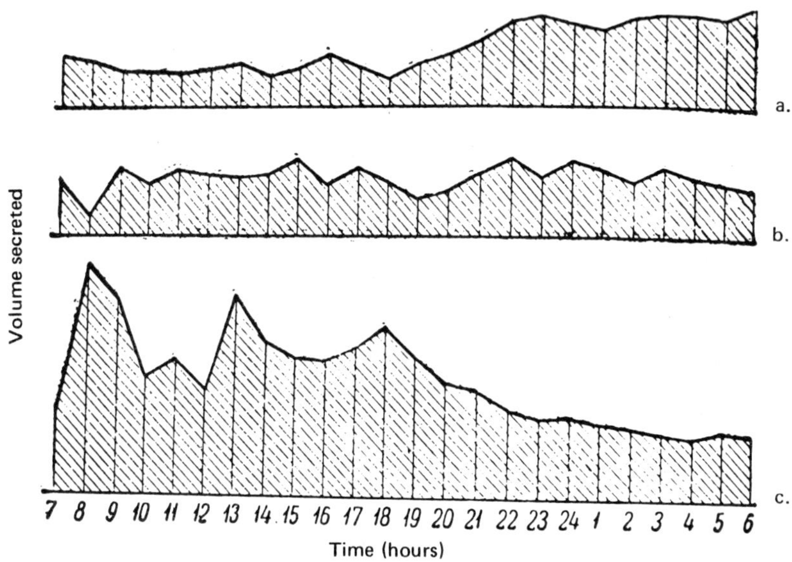

a. A piglet before weaning

b. A piglet after weaning

c. An adult pig

Fig. 5.5 Secretion of gastric juice through the day from a gastric pouch (Kvasnitskii, 1951)

PEPSIN

Lewis, Hartman, Liu, Baker & Catron (1957), Braude, Dollar, Mitchell & Porter (1958) and Hartman, Hays, Baker, Neagle & Catron (1961) measured pepsin levels in piglet gastric mucosa. They all found very low levels in the first few days of life, which increased slowly for the first two weeks, and then increased rapidly. Braude et al. and Hartman et al. also measured pepsin levels in the stomach contents, where they did not find a similar consistent rise. This could have been due to dilution of the gastric contents by varying quantities of food. Sève & Laplace (1975) found that the pepsin concentration in stomach contents of early weaned piglets increased with age over the age range 18–31 days if the sample was taken before feeding, but not if it was taken after a feed.

PANCREATIC SECRETION

The pancreas in mammals has two distinct functions. It is an endocrine organ, secreting hormones such as insulin and glucagon into the blood, and also an exocrine organ, secreting pancreatic juice through the pancreatic duct into the duodenum. It is primarily this exocrine secretion which is part of the digestive process. Pancreatic juice contains bicarbonate, and is the main factor reducing the acidity of the chyme entering the duodenum from the stomach, and bringing the pH to near neutrality. It also contains a range of hydrolytic enzymes acting on starch, proteins, fats and nucleic acids.

In the exocrine pancreatic cells there are small vacuoles containing granules which can be shown to consist mainly of precursors of the proteolytic enzymes. These granules provide temporary storage for the enzymes which can be released into the pancreatic ducts in response to stimuli (Palade, 1959; Palade, Siekevitz & Caro, 1962).

In the pig, the pancreatic duct usually joins the bile duct, the two secretions mixing before entering the duodenum by a common duct, but sometimes the pancreatic and bile ducts are separate. Vodovar, Flanzy & Francois (1964), in a study on over 100 pigs, found that, in 78% of the animals, the pancreatic and bile ducts joined before entry into the duodenum, while in the remaining 22% they entered the duodenum separately.

Composition

Pancreatic juice in the pig, as in other mammals, contains sodium and potassium in approximately the same concentrations as in plasma, and chloride and bicarbonate as the main anions, with low concentrations of other electrolytes (Hickson, 1970a). Table 5.3 gives the sodium, potassium and chloride and bicarbonate values obtained by Hickson.

The only organic components present in significant quantities are a range of enzymes and pro-enzymes listed in Table 5.4, and polypeptide enzyme co-factors and inhibitors (Table 5.5). The protein composition of pig pancreatic juice has been studied by Marchis–Mouren (1959, 1965) and Marchis–Mouren, Charles, Ben Abdeljlil & Desnuelle (1961) by separation on ion-exchange columns. They were able to identify almost every peak found in their fractionation with an enzyme or pro-enzyme, and showed that over 90% of the protein in pancreatic juice is enzyme or pro-enzyme. Uriel & Avrameas (1965), by immuno-electrophoresis of pig pancreatic extracts, were able to demonstrate a

Table 5.3. Concentration of principal inorganic ions in the pancreatic juice of the pig secreted at rates of 0.04–41.21 ul/g gland/min. in response to stimulation of the vagus nerves and injections of secretin. (Hickson, 1970a)

	Samples	Concentration (meq/l) (means ± 05% confidence limits)	No. of observations	No. of animals
1.	Pancreatic juice			
	Anions			
	($Cl^- + HCO_3^-$)			
	(i) Stimulation	148.6±4.6	39	5
	(ii) Secretin	157.4±2.9	66	6
	Cations Na^+ (all samples)	148.6±1.2	105	5
	K^+ (all samples)	5.7±0.2	105	5
2.	Plasma			
	Cations Na^+	142.3±8.2	9	8
	K^+	4.5±0.7	8	8

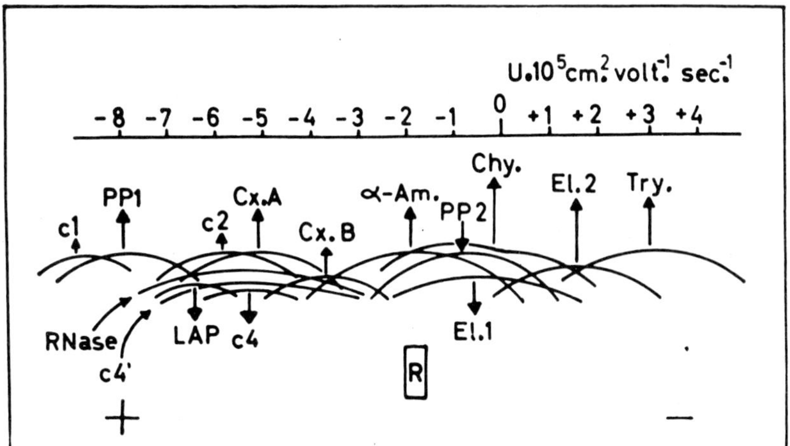

Abbreviations: PP1 and PP2 (pancreatic proteases 1 and 2); Cx.A and Cx.B (carboxypeptidase A and B); RNase (ribonuclease; LAP (leucine aminopeptidase); α-Am (α-amylase); El.1 and El.2 (elastases 1 and 2); Chy. (chymotrypsin); Try. (trypsin); c1, c2, c4 (antigenic constituents, c1, c2 and c4'); R (starting reservoir).

Fig. 5.6 Immunoelectrophoretic diagram of swine pancreatic extracts. Individual enzymes were located by the appropriate reactions in gel media (from Uriel and Avrameas, 1965. Reprinted with permission from 'Biochemistry,' 4, 1740, 1965. Copyright by the American Chemical Society).

Table 5.4 *Enzymes obtained from pig pancreatic juice*

Class	Name	E.C. (1972) number	Secreted as	Reference
Carbohydrase	α-amylase 1	3.2.1.1	enzyme	1
	α-amylase 2	3.2.1.1	enzyme	2
	Chitinase	3.2.1.14	enzyme	3
Protease				
(endopeptidase)	Trypsin	3.4.21.4	Trypsinogen	4, 5, 6
	Cationic trypsin	3.4.21.4	Trypsinogen	7, 8
	Chymotrypsin A	3.4.21.1	Chymotrypsinogen A	9, 10
	Chymotrypsin B	3.4.21.1	Chymotrypsinogen B	11
	Chymotrypsin C	3.4.21.2	Chymotrypsinogen C	11, 12, 13
	Elastase	3.4.21.11	Proelastase	14, 15, 16, 17
(exopeptidase)	Carboxypeptidase A	3.4.12.2	Procarboxypeptidase A	18, 19
	Carboxypeptidase B	3.4.12.3	Procarboxypeptidase B	20, 21
Lipase	Triacylglycerol lipase (Lipase)	3.1.1.3	enzyme	22
	Phospholipase A_2	3.1.1.4	Prophospholipase A_2	23, 24, 25, 26, 27
	Cholesterol esterase	3.1.1.13	enzyme	28, 29
Nuclease	Ribonuclease I	3.1.4.22	enzyme	30, 31, 32, 33

References (Table 5.4)

1. Meyer, Fischer & Bernfeld (1947)
2. Cozzone, Pasero, Beaupoil & Marchis-Mouren (1970)
3. Jeuniaux (1961)
4. Charles, Rovery. Guidoni & Desnuelle (1963)
5. Abita, Delaage, Lazdunski & Savdra (1969)
6. Smith & Liener (1967)
7. Louvard & Puigeserver (1974)
8. Voytek & Gjessing (1971)
9. Rovery, Charles, Guy, Guidoni & Desnuelle (1960)
10. Charles, Gratecos, Rovery & Desnuelle (1967)
11. Gratecos, Guy, Rovery & Desnuelle (1969)
12. Folk & Schirmer (1965)
13. Peanasky, Gratecos, Baratti & Rovery (1969)
14. Grant & Robbins (1957)
15. Lewis, Williams & Brink (1956)
16. Shotton & Hartley (1970)
17. Shotton & Watson (1970)
18. Folk & Schirmer (1963)
19. Folk (1963)
20. Folk, Piez, Caroll & Gladner (1960)
21. Roholt & Pressman (1971)
22. Verger, de Haas, Sarda & Desnuelle (1969)
23. de Haas, Postema, Niewenhuizen & van Deenan (1968)
24. de Haas, Postema & Niewenhuizen (1968)
25. Abita, Lazdunski, Bonsen, Pieterson & de Haas (1972)
26. de Haas, Slotboom, Bonsen, van Deenen, Maroux, Puigserver & Desnuelle (1970)
27. de Haas, Slotboom, Bonsen, Niewenhuizen, van Deenen, Maroux, Dlouha & Desnuelle (1970)
28. Swell, Field & Treadwell (1954)
29. Momsen & Brockman 1977)
30. Yamasaki, Murakami, Irie & Ukita (1968)
31. Jackson & Hirs (1970a)
32. Jackson & Hirs (1970b)
33. Phelan & Hirs (1970)

Table 5.5 Peptides in pig pancreatic juice which act as enzyme cofactors and inhibitors

Name	Amino acid residues	Function	References
Colipase I	87	Assist hydrolysis of insoluble triglycerides by triacyl-glycerol lipase	Charles, Erlanson, Bianchetta, Joffre, Guidoni & Rovery (1974)
Colipase II	84		Erlanson, Charles, Astier & Desnuelle (1974)
Trypsin inhibitor I	56	Reversibly bind with trypsin to give an inactive complex hydrolysed by excess trypsin	Tschetsche, Wachter, Kupfer & Niedermeier (1969)
Trypsin inhibitor II	52		

slightly larger number of enzymes (Fig. 5.6). While Table 5.4 lists the enzymes which have so far been clearly and unambiguously isolated and described, it is now clear that other enzymes occur. A number of enzymes have been isolated, partially characterized and given different names by different authors and some of these may be identical with enzymes which are now clearly defined. Inconsistency in the use of terms has tended to confuse the situation. For example, the term "elastase" has been used by some authors for any enzyme which hydrolyses elastin, but it is reserved by others for enzymes which preferentially hydrolyse amides or esters of alanine or similar neutral amino acids with little or no side-chain. At least one enzyme which has been called an elastase on account of its elastolytic activity, also has very high hydrolytic action on tyrosine amides and esters and could thus be classified as a chymotrypsin. Table 5.6 lists the possible correlation with the nomenclature we are using, of a number of enzymes isolated from pig pancreas or pancreatic juice. Some of these may not correspond with any enzymes which are as yet fully defined. The digestive properties of these enzymes are dealt with in the chapters on digestion of the relevant dietary constituents.

Because of the very considerable effect a number of stimuli have on secretion rate and composition of pancreatic juice, no definite values can be given. Hickson (1970a) found that in 4 to 12 week old pigs (8–20 kg) under anaesthesia, secretion varied from 0.39±0.12 μl/g gland/min without any stimulus to 30.3±4.3 μl/g gland/min with maximum stimulus.

Figures for the concentrations of some enzymes in individual samples of pig pancreatic juice given by Marchis–Mouren et al. (1961) illustrate the variation from one sample to another, which was also remarked on by Pekas, Thomson & Hays (1966) on the basis of studies on 3 to 7 week old pigs.

Control of pancreatic secretion

Kvasnitskii (1951) and Pekas (1965) found that pancreatic secretion was continuous in the pig. Kvasnitskii showed that although in all but the very young pig, pancreatic secretion varied during the day, increasing very much after feeding, it did not stop altogether in a fasted animal, a finding confirmed by Corring, Aumaitre & Rérat (1972). Kvasnitskii regarded this as due to the continuous nature of gastric secretion in the pig, producing a continuous stimulation of the pancreas, and reported that if gastric secretion was prevented from entering the duodenum, by washing out the stomach in a gastric fistularized pig, pancreatic secretion fell to zero. However, Hickson (1970a) demonstrated a slight but definite continuous secretion in the completely unstimulated pig pancreas. This almost complete absence of pancreatic secretion in the absence of any stimulus is also usual in other species. The presence of acid and food in the upper small intestine is the primary stimulus to pancreatic secretion, causing the release into the portal blood of peptide hormones from the intestinal mucosa, (see Chapter 4) and also causing vagal nervous stimulation of the pancreas. One of the hormones, secretin, released in response to acid in the duodenum, stimulates the secretion of fluid and bicarbonate, while the other, C.C.K., released in response to food in the duodenum, stimulates secretin of pancreatic enzymes. Vagal nervous stimulation of the pancreas has both effects.

In pancreatic tissue in vitro, the different enzymes are secreted in the same proportions under a variety of different conditions and stimuli (Scheele & Palade, 1975; Tartakoff, Jamieson, Scheele & Palade, 1975), so the immediate secretory stimulus appears to effect the release of all the enzymes equally.

Table 5.6 Possible correlation of further proteases isolated from pig pancreas with current enzyme classification

Authors	Name given	Probable identity
Hofstree (1952a, b)	Esterase 1	Chymotrypsin C
Grant & Robbins (1956)	Pankrin	Chymotrypsin C
Gjessing & Hartnett (1962)	Esteroproteolytic enzyme	Chymotrypsin C
Uriel & Avrameas (1965)	Protease 1	Chymotrypsin C
Baumstark (1970)	Protease	Chymotrypsin C
Avrameas & Uriel (1965)	Elastase 2	Elastase
Ardelt (1974, 1975)	Elastase 2	Intermediate category between chymotrypsins & elastase
Ledoux & Lamy (1975)	Elastase A	Elastase
Ledoux & Lamy (1975)	Elastase B	Elastase 2 (Ardelt, 1974)
Avrameas & Uriel (1965)	Elastase 1	
Uriel & Avrameas (1965)	Protease 2	Unidentified. May be elastase 2 of Ardelt (1974)
Uram & Lacy (1969)	Elastase	
Reinhardt & Uhlenbruck (1970)	Elastolytic enzyme	

Secretin

The presence of secretin in the blood causes the pancreas to secrete a greatly increased volume of fluid with a high bicarbonate concentration (Bayliss & Starling, 1902). The bicarbonate concentration rises with increasing secretion rate, while the chloride concentration falls, the sodium and potassium levels in the juice remaining constant. In the young pig, the bicarbonate concentration of the pancreatic juice rises with secretin stimulation from a resting value of 30–50 meq/1 to a maximum value of 130–160 meq/l, while the chloride concentration falls correspondingly from a resting value of 130 meq/l to a minimum of 18–24 meq/l (Fig. 5.7B, Hickson, 1970a).

The most powerful stimulus for release of secretin into the blood from the duodenal mucosa is a low pH in the duodenal lumen, though other stimuli, such as various food digestion products, can be effective. The volume of pancreatic juice secreted by an experimental pig with a duodenal loop has been shown to be directly related to the pH of the duodenal lumen over the range pH 2 to 7 (Magee & Hong, 1966; Hong, Nakamura & Magee, 1967) (Fig. 5.8).

Cholecystokinin (C.C.K.)

The presence of CCK in the blood does not alter the volume or electrolyte concentration of the pancreatic juice, but causes a large increase in the concentration of digestive enzymes in the juice (Harper & Raper, 1943; Jorpes & Mutt, 1966). This is due to the release into the juice of the zymogen granules which are stored in the acinar cells (Palade, 1959).

The hormone is released in response to the presence in the duodenal lumen of nutrients, and particularly fat digestion products. The difference in the pancreatic response to the presence in the duodenal lumen of different substances, was demonstrated in the pig by Hong & Magee (1970), hydrochloric acid producing predominantly a secretin response and oleic acid predominantly a CCK response (Table 5.7).

Table 5.7 *Effect of intraduodenal stimulants on pancreatic secretion. Each figure paired difference from control (from Hong & Magee, 1970).*

	Volume (ml)	Amylase (Arbitrary units)	Lipase	Protein (mg
Oleic acid	3.3±0.5	235±77	8.1±2.0	10.5±3.4
Peptone	2.8±0.8	283±73	9.2±3.0	6.1±1.9
HCl	5.3±0.7	382±83	8.3±1.9	11.3±3.2
Olive oil	0.3±0.3*	199±76	0.8±0.3	0.8±0.3
Amino acids	0.8±0.1	111±14.4		2.2±0.4
Alcohol	2.7±0.6	969±325	1.2±0.5	4.8±1.3

Each figure is the mean increase following stimulation ± the standard error of the mean. All the differences are significant at the 99% level except *. The number of animals making up each figure is from 20 to 9.

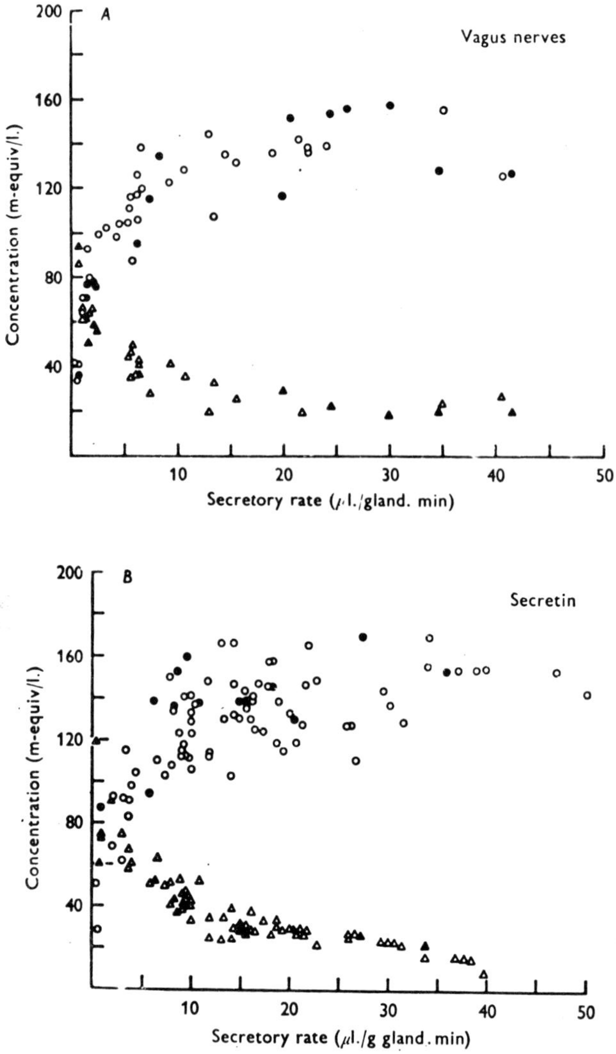

Fig. 5.7 *Concentration of bicarbonate (○) and chloride (Δ) ions in pancreatic juice of the pig at different rates of secretion. Flow of juice maintained in (A) by stimulation of the vagus nerves at selected frequencies (five pigs), in (B) by graded I.V. infusions of secretin (eight pigs). (○), (△) samples before, (●), (▲) samples after, atropine (0.5 mg/kg). (Hickson, 1970a).*

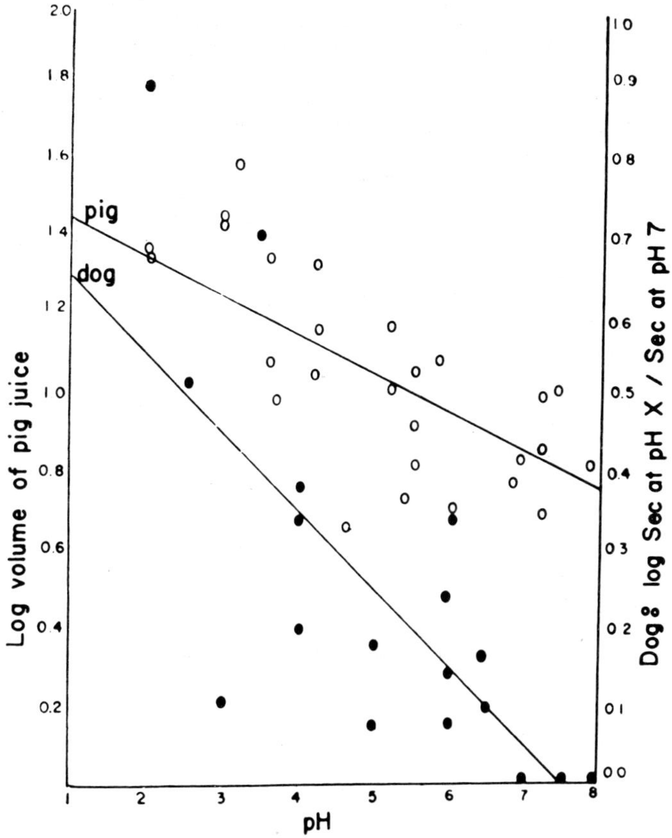

Fig. 5.8 The relationship between the pH of duodenal instillates, and the log. of the volume of pancreatic juice in pigs, and the log. of ratio between secretions at pH 7, and the secretion at the other pH's tested in dogs. Regression lines: pig r - 0.882; n - 30; dog r = 0.80; n = 24. (from Hong, Nakamura & Magee, 1967).

Vagal nerve stimulation

Stimulation of pancreatic secretion in the pig by the thoracic vagal nerves has been demonstrated by Hickson (1963), Magee & White (1965), and by Hickson (1970a) who showed that stimulation of the vagal nerves leading to the pancreas produced a copious flow of secretion rich in bicarbonate, amylase, lipase and proteolytic enzymes. The maximum flow rate was 30.3±4.3 μl/g gland, and during this stimulated secretion, the bicarbonate concentration rose from a resting level of 20–30 meq/l reaching a level of 130–160 meq/l at flow rates of 10–20 μl/g gland/min or more. During this time the chloride fell in a complementary way from a resting value of 130 meq/l to a level of 18–24 meq/l at flow rates of 10–20 μl/g gland/min or more (Fig. 5.7A). This effect could still be produced after resection of the stomach and small intestine, and was therefore independent of the action of hormones from the gastrointestinal tract.

Hickson (1970b) showed that in the pig, unlike the dog, the vagal stimulus is of more importance than the hormonal in pancreatic enzyme secretion.

Response to feeding

Kvasnitskii (1951) showed that the pancreatic secretion of unweaned piglets was fairly uniform throughout the day while that of older pigs varied considerably, with sharp responses to feeding (Fig. 5.9).

The increase in protein secretion by the pig pancreas in response to food was shown clearly by Corring, Aumaitre & Rérat (1972), who found two different types of response in secretory volume. In some individuals there was a large rise in secretory volume in response to a feed together with a slight rise in protein concentration of the juice (Fig. 5.10), while in others there was a slight fall in volume secreted, with a very large rise in protein concentration and, later, a moderate rise in volume (Fig. 5.11).

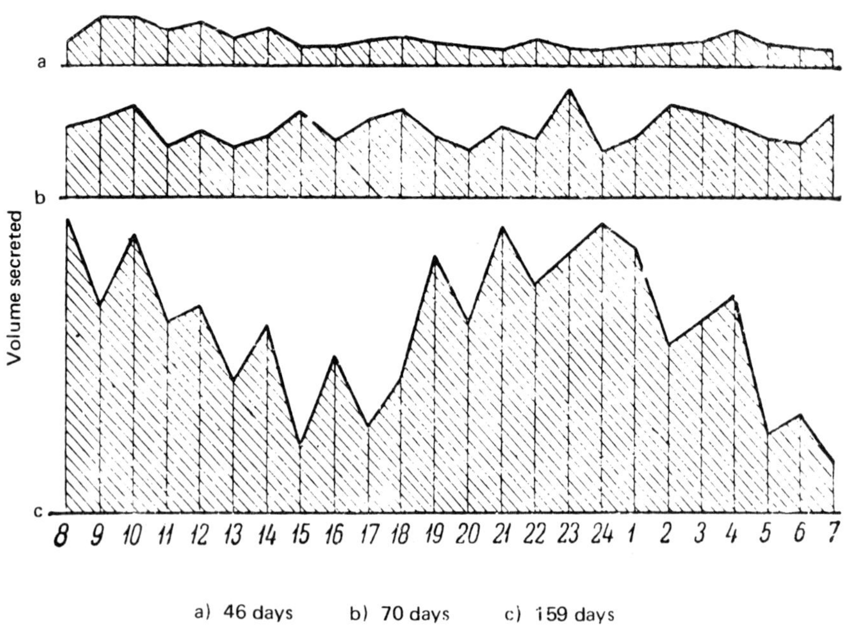

Fig. 5.9 Volume of pancreatic juice secreted through the day by a fistularized pig at various ages (Kvasnitskii, 1951).

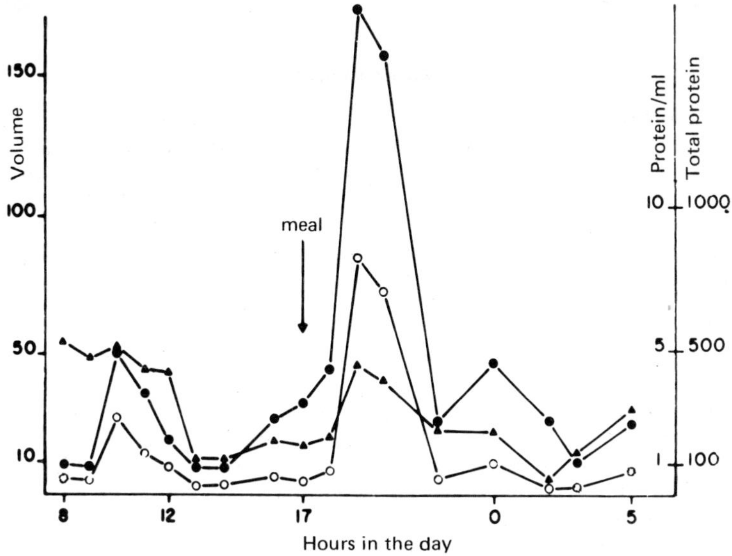

Fig. 5.10 Influence of feeding on pancreatic secretion in the pig. (1st type of response) (Corring, Aumaitre & Rérat, 1972).

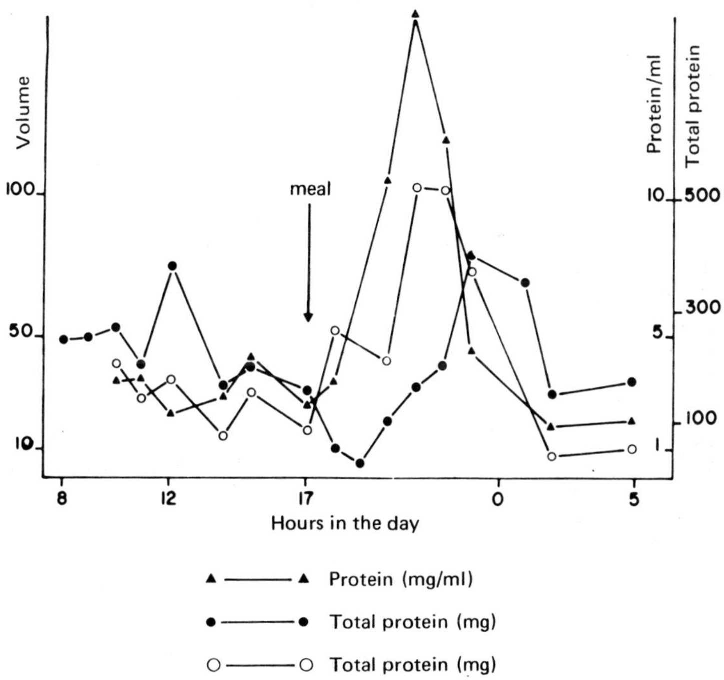

Fig. 5.11 Influence of feeding on pancreatic secretion in the pig. (2nd type of response) (Corring, Aumaitre & Rérat, 1972).

Variation with diet

The proportion of the different enzymes secreted by the pancreas can also be influenced by the diet. Aumaitre (1971) obtained low amylase and high chymotrypsin levels in the pancreatic juice of 5 to 6 week old piglets fed on a high-protein, low-starch diet and obtained the reverse by replacing the protein by starch. Corring & Saucier (1972) confirmed this in 40 kg pigs with re-entrant fistulas of the pancreatic duct, obtaining considerably higher chymotrypsin levels in the pancreatic juice when the pigs were fed high protein diets, the rise in chymotrypsin starting about 48 hours after raising the protein level of the diet (Fig. 5.12), while the trypsin level did not respond to dietary changes. This is in agreement with findings in other species. Corring (1975) did similar experiments in which low-starch, high-fat (arachis oil) and high-starch, low-fat diets were interchanged at weekly intervals. He obtained low amylase and high lipase levels in the pancreatic juice on the former diet and high amylase and low lipase levels on the latter, while the chymotrypsin and total protein concentrations remained constant. The response in composition of the pancreatic juice started about 3 days after the change of diet.

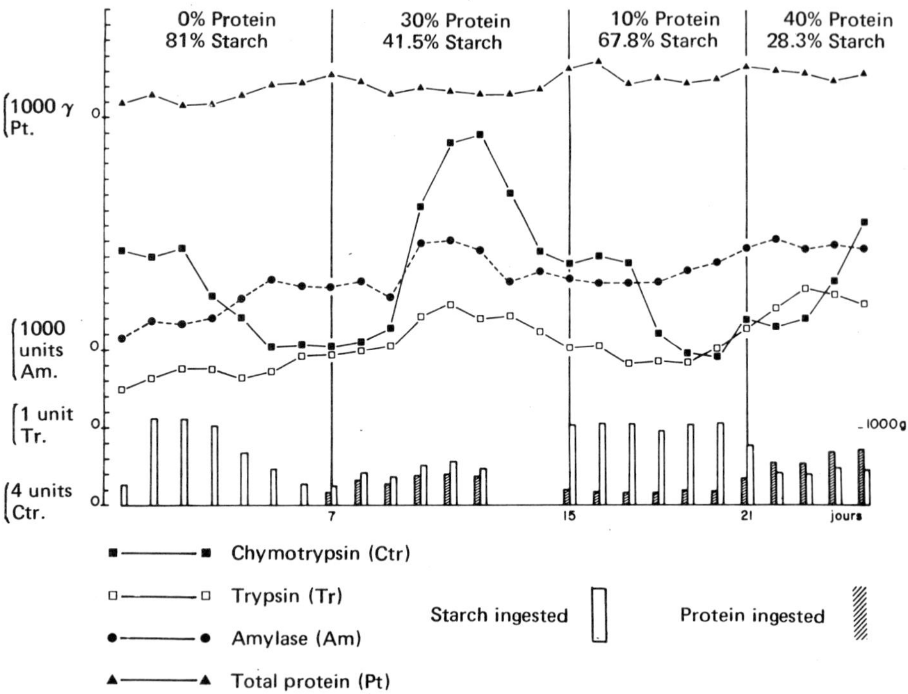

Fig. 5.12 Specific enzyme activities (activity per mg. protein) and total protein per ml in samples of pancreatic juice. Means per 24 hours (Corring & Saucier, 1972).

Feedback regulation

Hong, Nakamura & Magee (1967) showed that the introduction of pancreatic juice into the duodenum of the pig caused very marked suppression of secretion, and that a corresponding amount of either bicarbonate or boiled juice did not have a comparable effect. Corring (1972, 1974, 1975) showed that diversion of pancreatic secretion from the duodenum doubled the protein output and multiplied five-fold the volume of the secretion, and that re-introduction of the pancreatic juice into the duodenum reduced the secretion to its original level. Only re-introduction into the duodenum was effective, and the enzyme fraction of the juice seemed to be the relevant part. This is in agreement with the results obtained by Schneeman & Lyman (1975) in rats, who regarded active trypsin and chymotrypsin as the relevant factors, and considered that they acted by suppressing release of CCK by the mucosa.

Variation with age

Gorrill & Friend (1970) found that fresh pancreas weight in piglets increased linearly with body weight and Corring (1975) in a study of over 60 piglets from birth to 8 weeks, found an even closer linear relationship between log pancreas weight and log live weight.

Measuring the total quantities of the enzymes in the pancreas, Corring (1975) found that chymotrypsin (or chymotrypsinogen) was the earliest to appear in quantity, being present at significant levels from birth to 1 week and rising gradually. Lipase appeared in significant amounts at 1–2 weeks and increased rapidly to 6 weeks. Amylase was not present in any quantity till 3–4 weeks and trypsin (or trypsinogen) at 4 weeks, both of these then increasing rapidly with age. In general these results agree with the smaller earlier studies of Kitts, Bailey & Wood (1956) and Hartman, Hays, Baker, Neagle & Catron (1961) on amylase and lipase and Gorrill and Friend (1970) on trypsin and chymotrypsin.

Kvasnitskii (1951) showed that the volume of pancreatic juice secreted by a pig increased rapidly with age over the first few months of life rising more or less linearly with age from 500 mls per day at 5–6 weeks to over 8 litres at 7–8 months.

Variation with conditions

Szabó, Ribiczeyné–Szabó & Rafai (1976) reported that the amylase and lipase levels in the pancreas of 4–5 week old piglets were influenced by environmental temperature, being high in piglets kept at lower temperatures and low in pigs kept at high temperatures.

Sullivan, Birch, Quigley & Magee (1974) found decreased pancreatic output and decreased numbers of granules in the pancreas of zinc-deficient pigs. They did not find a decrease in the cellular zinc-containing enzymes but did not report on the digestive zinc-containing enzymes (the carboxypeptidases).

BILE

Bile is secreted by the polygonal cells of the liver into the bile canaliculi which lead into the hepatic bile duct. In the pig, as in other mammals, the secretory process involves primarily the plasma membrane of the cell, so that plasma membrane enzymes are found in bile, while intracellular enzymes are not (Holdsworth & Coleman, 1975).

In most mammalian species, including the pig, bile passes into the gall bladder, an extensible blind side-branch of the bile duct, where it is stored and concentrated. When the gall bladder contracts in response to the hormone CCK, this bile is expelled down the common bile duct, through the sphincter of Oddi, into the duodenum.

Composition

Egger, Kutz, Strebel, Bircher, Weber, Scholl & Preisig (1974) measured the levels of sodium, potassium, chloride, bicarbonate, bile salt, and the pH of bile from bile-duct cannulated pigs while bile flow was stimulated by bile duct infusion. They found sodium levels of 165–196 meq/l, potassium 3.9–5.9 meq/l, chloride 73–97 meq/l and bicarbonate 11.3–26.5 meq/l. The pH varied from 7.43 to 7.91. In each sample the excess of the two cations over the two anions corresponded closely with the bile salt concentration.

The principal organic constituents of bile are bile salts, phospholipids, cholesterol, mucus and excretory waste products, especially bile pigments. The components relevant to digestive function are the bile salts and to a lesser extent, the phospholipids.

Bile salts

The characteristic feature of bile is the presence of highly surface-active derivatives of sterols. In mammals these substances are conjugated bile acids, which consist of a steroid carboxylic acid conjugated with either glycine or taurine. These are largely ionised at the pH of bile.

The conjugated bile acids in the lumen of the small intestine assist in the emulsification of the fat, and particularly in the micellar dispersion of the products of fat digestion, facilitating their absorption. The bacteria in the lower small intestine deconjugate the bile acids and to some extent alter them chemically, mainly by reduction of some of the hydroxyl groups. The altered and unaltered bile acids are reabsorbed in the lower small intestine, pass into the portal blood and are absorbed by the liver. Together with some newly-formed bile acid they are then conjugated and again secreted into the bile.

From the published lists of bile acids whose conjugates are found in pig bile (Lehmann, Tepper & Hilgetag, 1964; Haslewood, 1967; Matschiner, 1971), a composite list of 7 bile acids can be made (Fig. 5.13). Of these, the main constituents are hyodeoxycholic, hyocholic and chenodeoxycholic acids (Bergström, Danielsson & Göransson, 1959) mainly conjugated with glycine, but also to some extent with taurine (Haslewood & Sjövall, 1954).

Bergström et al. (1959) showed that hyocholic acid was formed in the pig liver from chenodeoxycholic acid and that the hyodeoxycholic acid originated from hyocholic acid which had been reduced in the intestine lumen and re-cycled. The other three bile acids present in traces probably also originate from intestinal reduction of the main constituents. Thus the pig liver, like the liver of many other species, first produces chenodeoxycholic acid. Part of this, and part of the chenodeoxycholic acid which has been recycled from the intestine, is hydroxylated to hyocholic acid, a substance peculiar to the bile of suidae

Major constituents	(Ref. no)
Hyodeoxycholic acid	(1, 2)

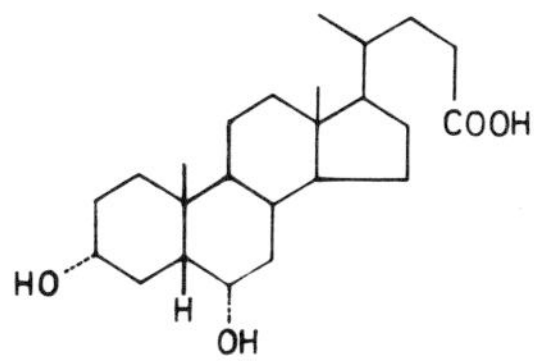

Hyocholic acid	(3, 4)

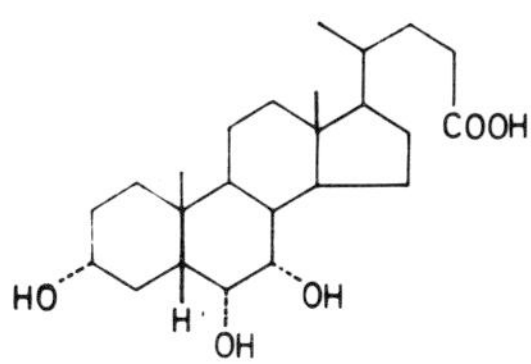

Chenodeoxycholic acid	(5, 6)

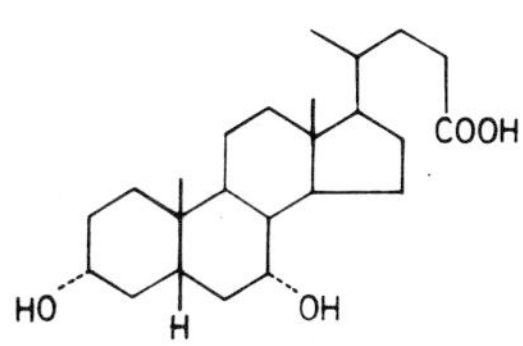

References
1. Gundelach & Strecker (1847)
2. Windaus & Bohne (1923)
3. Haslewood (1956)
4. Ziegler (1956)
5. Ido & Sakurai (1939)
6. Lehmann, Koppe, Tepper & Hilgetag (1964)
7. Matschiner et al. (1958)
8. Kimura (1937)
9. Shimizu (1953)
10. Fernholz (1935)
11. Sugiyama (1937)

Minor constituents	(Ref. no)
3α,6β-dihydroxycholanic acid	(7)

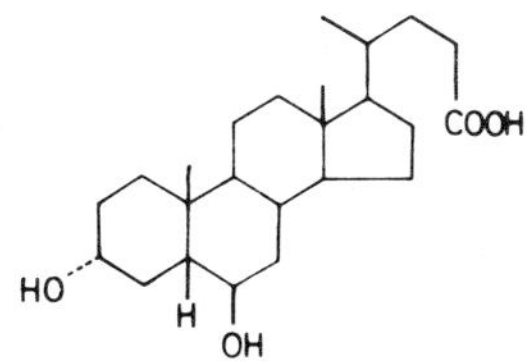

3β,6α-dihydroxycholanic acid	(8)

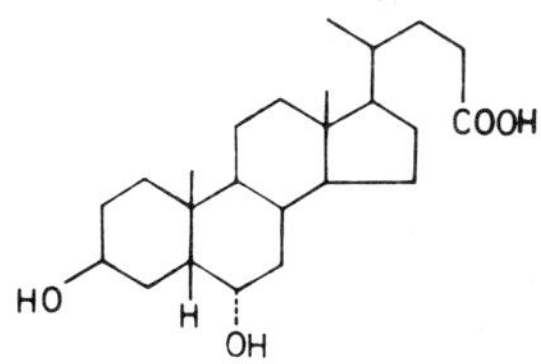

Lithocholic acid	(9)

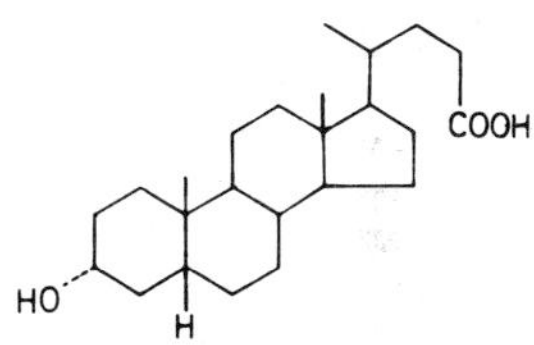

3α-hydroxy-6-keto-allocholanic acid	(6, 10, 11)

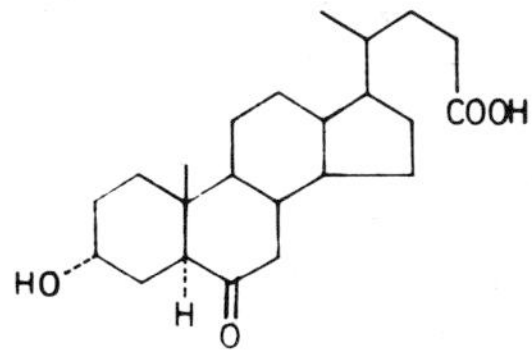

Fig. 5.13 Structure of bile acids isolated from pig bile.

(Matschiner, 1971). Cholic acid, a major constituent of the bile of most mammals, is absent from pig bile, or present only in traces (Haslewood, 1967; Matschiner, 1971), its place being taken by hyocholic and hyodeoxycholic acids. It is not known exactly what the relative benefits of having a particular bile acid are to a species, but the different acids have slightly different dissociation constants and so will have maximum micelle-forming activity over slightly different pH ranges. They may also act optimally with slightly different mixtures of fatty acids and monoglycerides.

Biliary phospholipids

Bile contains lecithin which is hydrolysed in the intestine by pancreatic phospholipase to lysolecithin, a substance which contributes to the micellar dispersion of the products of fat digestion.

Control of secretion

Bile secretion is continuous, but not uniform, being greatly increased by the presence in the blood of bile salts, secretin, or CCK.

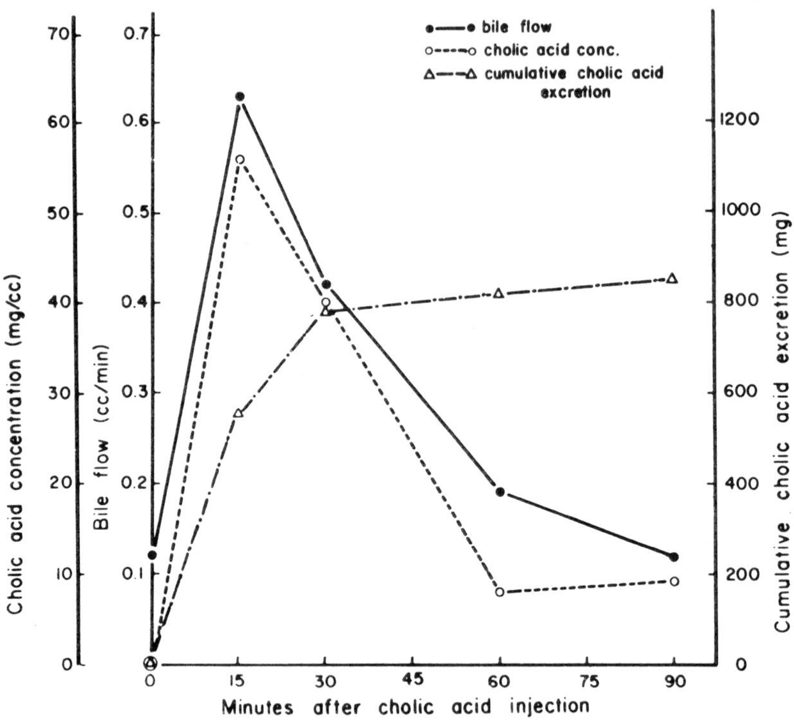

Fig. 5.14 Changes in bile flow and bile salt concentration following injection of 1 g of sodium cholate. Bile flow increased to 5 times basal rate concomitant with the excretion of a major portion of the injected load of sodium cholate (Hardison & Norman, 1967).

Bile salts

Bile salts produce an increase in biliary fluid output while not having any significant effect on electrolyte composition. This effect, which has long been known in man and the dog, was demonstrated in the perfused pig liver by Hardison & Norman (1967) (Figs. 5.14 & 5.15). Egger et al. (1974) showed the same effect by perfusing taurocholate intravenously into 29–42 kg pigs with bile duct cannulas. As the infusion rate of bile salts was increased, so the bile output increased so that the composition of the bile with respect to the components measured (sodium, potassium, chloride, bicarbonate, bile salts and pH), did not alter significantly. There appeared to be a basal secretion of 0.15 ml/min independent of bile salt, and beyond this secretion increased linearly with bile salt output, each microequivalent of bile salt secreted being associated with the secretion of 0.01 ml water, 1.87 μeq of sodium, 0.06 μeq of potassium, 0.59 μeq of chloride and 0.24 μeq of bicarbonate.

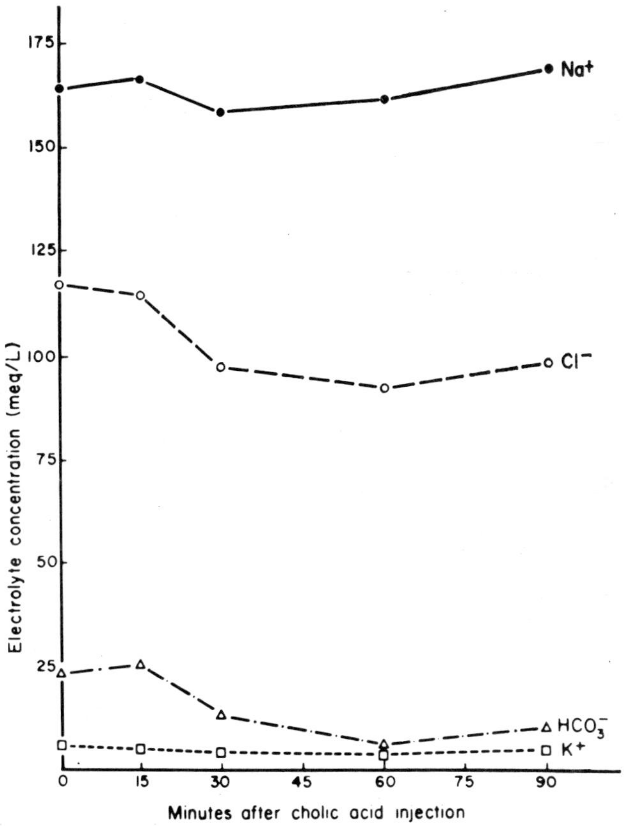

Fig. 5.15 Electrolyte composition of bile following injection of 1 g of sodium cholate. Chloride and bicarbonate concentrations both drop slightly. Cation concentrations are less affected (Hardison & Norman, 1967).

Secretin

As with the pancreas and the Brunner's glands, secretin stimulates the secretion of water and bicarbonate so that the chloride concentration of the bile falls as the bicarbonate concentration rises, the sodium concentration remaining constant, while that of the organic constituents falls. This effect, well known in man and the dog, has been demonstrated in perfused pig livers by Hardison & Norman (1967) (Fig. 5.16), Douglas, Watts, Jablonski & Owen (1969) and Gordon, Douglas, Jablonski, Sali & Watts (1972).

Hardison & Norman (1968) showed in secretin-stimulated perfused pig liver that, although the additional secretion over and above the basal level increased with increasing dose of secretin, the calculated concentration of this additional secretion was constant and independent of the secretin dose.

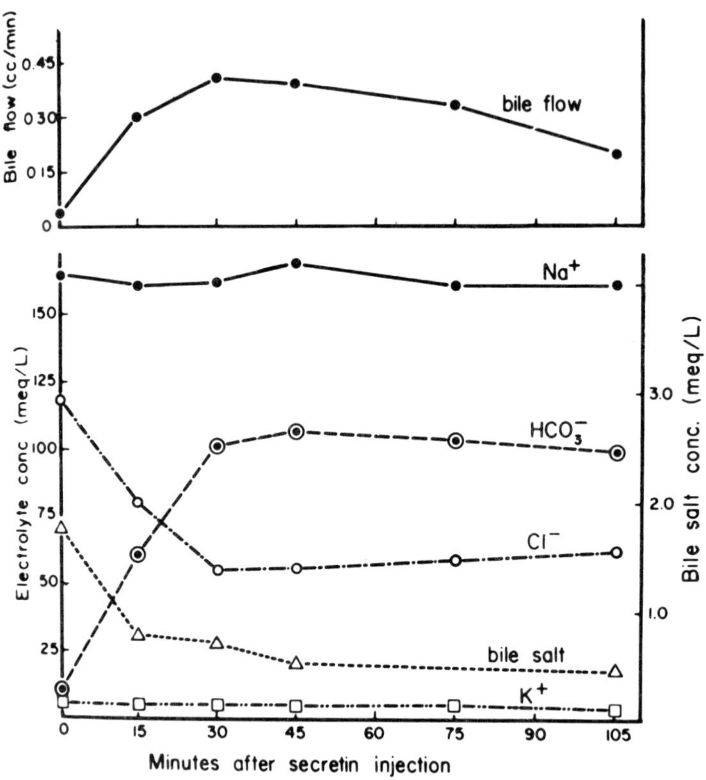

Fig. 5.16 Sequential changes in bile flow and electrolyte concentrations after injection of 100 units of secretin. As bile flow increases, bicarbonate concentration rises and chloride concentration falls. Bile salt concentration, low before injection, declines further after injection. (Hardison & Norman, 1967).

Cholecystokinin (CCK)

Gordon et al. (1972) showed that, in the isolated, perfused pig liver, CCK caused increased bile output with a raised bicarbonate concentration, but that the bile acid output did not alter, so that its concentration fell.

Emptying

The hormone CCK, released from the blood in response to the presence of fat digestion products in the duodenum (see gastrointestinal hormones), stimulates contraction of the gall bladder, relaxation of the sphincter of Oddi and so the passage of gall bladder bile into the duodenum (Ivy & Oldberg, 1928; Hong, Magee & Crewdson, 1956; Jorpes & Mutt, 1966).

Laplace (1976a, b) carried out electrical and pressure studies in the gall-bladder, the bile duct, duodenum, pylorus and gastric antrum of the pig. He found no evidence of active emptying by the gall-bladder but concluded that the emptying was passive, due to pressure differences and was controlled by the musculature at the sphincter of Oddi. This was duodenal musculature, the sphincter having no significant musculature of its own, and its contractions were related to those of the duodenum. Rhythmic contractions of the bile duct were recorded but these would only have been powerful enough to cause emptying if there were no resistance at the sphincter, and he did not regard them as being of importance in emptying bile into the duodenum under physiological conditions.

SMALL INTESTINE WALL

Brunner's glands

The duodenal mucosa is glandular, and from this tissue, known as Brunner's glands, there is a well-defined secretion. This was studied in various mammals including a 3-month old pig by Florey & Harding (1934). The duodenal secretion from the pig was similar to that of the other species, being viscous and alkaline, with a pH of 8.4–8.9 and alkalinity of 33 meq/l. The rate of secretion increased on feeding. In studies on a number of species including several 12 kg pigs, Wright, Jennings, Florey & Lium (1940) showed that duodenal secretion in a fasting pig was extremely slight but that, on feeding, it increased to 15.8–17.3 ml/h in an innervated duodenal pouch, but only to 9.8–13 ml/h in a denervated one. The alkalinity was 31 meq/l from the innervated and 23 meq/l from the denervated one. Florey & Harding (1934) showed an increase in water and bicarbonate secretion from the duodenum in response to secretin in the blood. Thus the duodenal secretion, like the gastric and the pancreatic, is mediated by a combination of nervous and hormonal stimuli.

Wright, Jennings, Florey & Lium (1940) demonstrated the absence of proteases, peptidases and lipase from the juice but found enterokinase and amylase. The amylase was detected by production of reducing sugar from starch, and so could have been glucoamylase, which is present in significant quantities in duodenal mucosa. Lysozyme was also found at a high level in the juice from the pig, but not that of the other species examined.

Florey, Jennings, Jennings & O'Connor (1939), in transposition experiments with pig duodenum, showed that the Brunner's gland secretion protected the intestinal wall from gastric contents which otherwise caused ulcers and perforation. It protects the duodenal wall from the very acid stomach contents by lining the duodenum with an alkaline viscous secretion. This, together with the alkaline bile and pancreatic juice helps to neutralize the stomach contents and so protects the regions further down the intestine.

The jejunal mucosa

There is net secretion of fluid into the lumen in the proximal half of the small intestine. This has been examined quantitatively by Horszczaruk, Buraczewska & Buraczewski (1974) who studied 3 pigs with two re-entrant fistulas, one at each end of a jejunal loop. This loop could be temporarily isolated from the flow of digesta and the fluid secreted into it collected, measured and analysed. Secretion was studied with the pig on a normal and a protein-free diet. The volume secreted into the loop varied considerably but averaged 300–400 ml per metre per 24 hours. The juice secreted on a normal diet contained 2–3% dry matter, 0.5–1.0% protein and 0.78–0.82% ash.

REFERENCES

Abita, J.P., Delaage, M., Lazdunski, M. & Savdra, J. (1969) The mechanism of activation of trypsinogen. The role of the four N-terminal aspartyl residues. *Eur. J. Biochem. 8,* 314–324.

Abita, J.P., Lazdunski, M., Bonsen, P.P.M., Pieterson, W.A. & de Haas, G.H. (1972) Zymogen enzyme transformations. On the mechanism of activation of phospholipase. *Eur. J. Biochem. 30,* 37–47.

Antonio, C.B. (1940) Studying on the strength of rennet extract produced from the stomach of pigs. *Philip. Agric., 29,* 518–526.

Akhipovets, A.I. (1956) O vozrastnykh osobennostyakh slyunootdeleniya u porosyat. *Sechenov fiziol. Zh. S.S.S.R., 42,* 882–886.

Ardelt, W. (1974) Partial purification and properties of porcine pancreatic elastase II. *Biochem. biophys. Acta, 341,* 318–326.

Ardelt, W. (1975) Physical parameters and chemical composition of porcine pancreatic elastase II. *Biochim. biophys Acta, 393,* 267–273.

Avrameas, S. & Uriel, J. (1965) Systematic fractionation of swine pancreatic hydrolases. II. Fractionation of enzymes insoluble in $(NH_4)_2SO_4$ solution at 0.40 saturation. *Biochemistry, N.Y., 4,* 1750–1757.

Aumaitre, A. (1971) Le developpement des enzymes dans le tube digestif du jeune porcelet[Importance pour le sevrage et signification nutritionelle. *Ann. Zootech., 20,* 551–575.

Bakeeva, E.N. (1962) Pishchevareniya i obmen veshchestv u molodnyaka svinei pri kachestvenno razlichnom proteinovum pitanii. *Nauchnye Trudy poltavskii nauchno-issledovatel'skii Institut Svinovodstva 22,* 18–35.

Bakeeva, E.N. & Utekhin, B.P. (1958) Functional state of the gastric glands in relation to their secretory activity in pigs. *Sechenov fiziol. Zh. SSSR, 44,* 1077–1081.

Baumstark, J.S. (1970) Isolation and partial characterisation of an elastase associated with acidic endopeptidase and its interaction with elastase at low ionic strength. *Biochim. biophys. Acta, 220,* 534–554.

Bayliss, W.M. & Starling, E.H. (1902) The mechanism of pancreatic secretion. *J. Physiol. Lond., 28,* 325–351.

Bergström, S., Danielsson, H. & Göransson, A. (1959) On the bile acid metabolism in the pig. *Acta chem. scand., 13,* 776–783.

Bovey, F.A. & Yanari, S.S. (1960) Pepsin. *in* The enzymes, Edn. 2, vol. 4. Eds Boyer, P.D., Lardy, H. & Myrbäck, K. pp. 63–91. Academic Press, New York.

Braude, R., Dollar, A.M., Mitchell, K.G. & Porter, J.W.G. (1958) Proteolytic enzymes and the clotting of milk in the stomach of the young pig. *Proc. Nutr. Soc., 17,* xlix.

Charles, M., Erlanson, C., Bianchetta, J., Joffre, J., Guidoni, A. & Rovery, M. (1974) The primary structure of porcine colipase II. I. The amino acid sequence. *Biochim. biophys. Acta, 359,* 186–197.

Charles, M., Gratecos, D., Rovery, M. & Desnuelle, P. (1967) Le chymotrypsinogène de porc. Purification et études de quelques propriétés *Biochim. biophys. Acta 140,* 395–409.

Charles, M., Rovery, M., Guidoni, A. & Desnuelle, P. (1963) Sur le trypsinogène et la trypsine de porc. *Biochim. biophys. Acta, 69,* 115–129.

Cheredkova, A.N. (1959) K voprosy o mekhanizme zheludochnogo sokotdeleniya u svinei. *Sbornik Rabot Leningradskii veterinarnyi Institut, 22,* 311–318.

Corring, T. (1972) Mécanisme de la sécretion pancréatique exocrine chez le porc: régulation par rétroinhibition. *Annls Biol. anim. Biochim. Biophys., 13,* 755–756.

Corring, T. (1974) Regulation de la secretion pancréatique par rétroaction négative chez le porc. *Annls Biol. anim. Biochim. Biophys., 14,* 487–498.

Corring, T. (1975) Adaptation de la secrétion du pancréas exocrine au régime alimentaire chez le porc, physiologie comparée, étude experimentale. Thesis, University of Paris, VI.

Corring, T., Aumaitre, A. & Rérat, A. (1972) Fistulation permanente du pancréas exocrine chez le porc. Application: réponse de la secretion pancréatique au repas. *Annls Biol. anim. Biochim. Biophys., 12,* 109–124.

Corring, T. & Saucier, R. (1972) Sécrétion pancréatique sur porc fistulé. Adaptation a la teneur en protéines du régime. *Annls Biol. anim. Biochim. Biophys., 12,* 233–241.

Cozzone, P., Paséro, L., Beaupoil, B. & Marchis-Mouren, G. (1970) Characterization of porcine pancreatic isoamylases: chemical and physical studies. *Biochim. biophys. Acta, 207,* 490–504.

Cranwell, P.D. (1978) Acid and pepsin secretion in young pigs reared solely by the sow or supplemented by solid food and weaned at 21 days. *Proc. Nutr. Soc., 36,* 142A.

Cranwell, P.D., Noakes, D.E. & Hill, K.J. (1968) Observations on the stomach contents of the sucking pig. *Proc. Nutr., Soc., 27,* 26A.

Cranwell, P.D., Noakes, D.E. & Hill, K.J. (1976) Gastric secretion and fermentation in the young suckling pig. *Br. J. Nutr., 36,* 71–86.

Cranwell, P.D., & Titchen, D.A. (1974) Gastric acid secretion in newly-born piglets. *Res. vet. Sci., 16,* 105–107.

Douglas, M.C., Watts, M. McK., Jablonski, P. & Owen, J.A. (1969) The effect of various gastroduodenal hormones on output of hepatic bile in the isolated perfused porcine liver. *Gut, 10,* 79.

Egger, G., Kutz, K., Strebel, H., Bircher, J., Weber, M., Scholl, E. & Preisig, R. (1974) Bile formation in the intact pig. *Am. J. vet. Res., 35,* 1203–1208.

Erlanson, C., Charles, M., Astier, M. & Desnuelle, P. (1974) The primary structure of porcine colipase II. II. The disulphide bridges. *Biochim. biophys. Acta, 359,* 198–203.

Fernholz, E. (1935) Die Isoliering der 3-Oxo-6-keto-allocholansäure aus Schweinegalle. *Hoppe-Seylers Z. Physiol. Chem., 232,* 202–205.

Florey, H.W. & Harding, H.E. (1934) Further observations on the secretion of Brunner's glands. *J. Path. Bact., 39,* 255–276.

Florey, H.W., Jennings, M.A., Jennings, D.A. & O'Connor, R.C. (1939) The reaction of the intestine of the pig to gastric juice. *J. Path. Bact., 49,* 105–123.

Folk, J.E. (1963) The porcine pancreatic carboxypeptidase A system. II. Mechanism of the conversion of carboxypeptidase A_1 to carboxypeptidase A_2. *J. biol. Chem., 238,* 3895–3898.

Folk, J.E., Piez, K.A., Caroll, W.R. & Gladner, J.A. (1960) Carboxypeptidase B. IV. Purification and characterization of the porcine enzyme. *J. biol. Chem., 235,* 2272–2277.

Folk, J.E. & Schirmer, E.W. (1963) The porcine pancreatic carboxypeptidase A system. I. Three forms of the active enzyme. *J. biol. Chem., 238,* 3884–3894.

Folk, J.E. & Schirmer, E.W. (1965) Chymotrypsin C. 1. Isolation of zymogen and active enzyme: Preliminary structure and specific studies. *J. biol. Chem., 240,* 181–192.

Gjessing, E.C. & Hartnett, J.C. (1962) Crystallization and characterization of an esteroproteolytic enzyme from porcine pancreas. *J. biol. Chem., 237,* 2201–2206.

Gordon, E.M., Douglas, M.C., Jablonski, P., Owen, J.A., Sali, A. & Watts, McK. (1972) Gastroduodenal hormones and bile-secretion studies in the isolated perfused pig liver. *Surgery, 72,* 708–721.

Gorrill, A.D.C. & Friend, D.W. (1970) Pancreas size and trypsin and chymotrypsin activities in the pancreas and intestinal contents of pigs from birth to 5 weeks of age. *Can J. Physiol., 48,* 745–750.

Grant, N.H. & Robbins, K.C. (1956) Partial purification and characterization of Pankrin, a new pancreatic proteinase. *J. Am. Chem., Soc., 78,* 5888–5890.
Gratecos, D., Guy, O., Rovery, M. & Desnuelle, P. (1969) On the two anionic chymotrypsinogens of porcine pancreas. *Biochim. biophys, Acta, 175,* 82–96.
Gregory, R.A. (1962) Secretory mechanisms of the gastro-intestinal tract. Monographs of the Physiological Society, Edward Arnold, London.
Gridin, N.Ya. (1956) K dinamike zheludochnogo sokotdeleniya u svinei na otdel'nye vidy kormov. *Sechenov fiziol. Zh. S.S.S.R., 42,* 773–778.
Gundelach, C. & Strecker, A. (1847) Untersuchung der Schweinegalle. *Justus Liebigs Annln Chem., 62,* 205–232.
de Haas, G.H., Postema, N.M. & Nieuwenhuizen, W. (1968) Purification and properties of an anionic zymogen of phospholipase A from porcine pancreas. *Biochim. biophys. Acta, 159,* 118–129.
de Haas, G.H., Postema, N.M., Nieuwenhuizen, W. & van Deenen, L.L.M. (1968) Purification and properties of phospholipase A from porcine pancreas. *Biochim. biophys. Acta, 159,* 103–117.
de Haas, G.H., Slotboom, A.J., Bonsen, P.P.M., Nieuwenhuizen, W., van Deenen, L.L.M., Maroux, S., Dlouha, V. & Desnuelle, P. (1970) Studies on phospholipase A and its zymogen from porcine pancreas. II. The assignment of the position of the six disulphide bridges. *Biochim. biophys. Acta, 221,* 54–61.
de Haas, G.H., Slotboom, A.J., Bonsen, P.P.M., van Deenen, L.L.M., Maroux, S., Puigserver, A. & Desnuelle, P. (1970) Studies on phospholipase A and its zymogen from porcine pancreas. I. The complete amino acid sequence. *Biochim. biophys, Acta, 221,* 31–53.
Hardison, W.G. & Norman, J.C. (1967) Effect of bile salt and secretin upon bile flow from the isolated perfused pig liver. *Gastroenterology, 53,* 412–417.
Hardison, W.G. & Norman, J.C. (1968) Electrolyte composition of the secretin fraction of bile from the perfused pig liver. *Am. J. Physiol., 214,* 758–763.
Harper, A.A. & Raper, H.S. (1943) Pancreozymin, a stimulant of the secretion of pancreatic enzymes in extracts of the small intestine. *J. Physiol, Lond., 102,* 115–125.
Hartman, P.A., Hays, V.W., Baker, R.O., Neagle, L.H. & Catron, D.V. (1961) Digestive enzyme development in the young pig. *J. Anim. Sci., 20,* 114–123.
Haslewood, G.A.D. (1956) Comparative studies of 'bile salts'. 9. The isolàtion and chemistry of hyocholic acid. *Biochem. J., 62,* 637–645.
Haslewood, G.A.D. (1967) Bile salts. Methuen, London.
Haslewood, G.A.D. & Sjövall, J. (1954) Comparative studies of bile salts. 8. Preliminary examination of bile salts by paper chromatography. *Biochem. J., 57,* 126–130.
Herriott, R.M. (1938) Isolation, crystallization and properties of swine pepsinogen. *J. gen. Physiol., 21,* 501–540.
Hickson, J.C.D. (1963) The effect of stimulation of the vagus nerves on pancreatic secretions in the pig. *J. Physiol. Lond., 168,* 24P–25P.
Hickson, J.C.D. (1970a) The secretion of pancreatic juice in response to stimulation of the vagus nerves of the pig. *. Physiol. Lond., 206,* 275–297.
Hickson, J.C.D. (1970b) The secretory and vascular response to nervous and hormonal stimulation in the pancreas of the pig. *J. Physiol., Lond., 206,* 299–322.
Hofstree, B.H.J. (1952a) Specificity of esterases. I. Identification of two pancreatic aliesterases. *J. biol. Chem., 199,* 357–364.
Hofstree, B.H.J. (1952b) Specificity of esterases. II. Behaviour of pancreatic esterases 1 & 2 towards a homologous series of n-fatty acid esters. *J. biol. Chem., 199,* 365–371.
Holdsworth, G. & Coleman, R. (1975) Enzyme profiles of mammalian bile. *Biochim. biophys. Acta, 839,* 47–50.
Höller, H. (1970a) Untersuchungen über Sekret und Sekretion der Cardiadrüsenzone im Magen des Schweines. I. Sekretionsvolumina und -rhythmik, Eigenschaften der Sekrete, *Zentbl. VetMed., 17A,* 685–711.
Höller, H. (1970b) Untersuchungen über Sekret und Sekretion der Cardiadrüsenzone im Magen des Schweines. II. Versuche zur Beeinflüssung der Spontansekretion isolierten Cardiadrüsenzone, Flüssigkeits- und Elektrolytsekretion in den mit verschiedenen Flüssigkeiten gefüllten isolierten kleinen Magen. *Zentbl. VetMed., 17A,* 857–873.
Hong, S.S. & Magee, D.F. (1970) Pharmacological studies on the regulation of pancreatic secretion in pigs. *Annls Surg., 172,* 41–48.
Hong, S.S., Magee, D.F. & Crewdson, F. (1956) The physiology of gallbladder evacuation. *Gastroenterology, 30,* 625–630.

Hong, S.S., Nakamura, M. & Magee, D.F. (1967) Relationship between duodenal pH and pancreatic secretion in dogs and pigs. *Annls Surg., 166,* 778–782.
Horszczaruk, F., Buraczewska, L. & Buraczewski, S. (1974) Ilość i sklad soku jelitowego wydzielanego do izolowanej petli jelita cienkiego u świń. *Roczn. Nauk roln., 95–B–4,* 69–77.
Huang, M.Z. & Liu, C.S. (1964) The influence of histamine on gastric pepsin secretion in the pig. *Acta physiol. Sinica, 27,* 140–144.
Ido, T. & Sakurai, R. (1939) Über das Vorkommen der Anthropodesoxycholsäure in der Galle des Schweins. *J. Biochem., Tokyo, 29,* 51–55.
Ivy, A.C. & Oldberg, E.A. (1928) A hormone mechanism for gallbladder contraction and evacuation. *Am. J. Physiol., 86,* 599–613.
Jackson, R.L. & Hirs, C.H.W. (1970a) The primary structure of porcine pancreatic ribonuclease. I. The distribution and sites of carbohydrate attachment. *J. biol. Chem., 245,* 626–636.
Jackson, R.L. & Hirs, C.H.W. (1970b) The primary structure of porcine pancreatic ribonuclease. II. The amino-acid sequence of the reduced S-aminoethylated protein. *J. biol. Chem., 245,* 637–652.
James, A.H. (1957) Physiology of gastric digestion. Monographs of the Physiological Society, Edward Arnold, London.
Jeuniaux, C. (1961) Digestion de la chitine chez les oiseaux et les mammifères. *Annls Soc. r. zool. Belg., 92,* 27–45.
Jorpes, J.E. & Mutt, V. (1966) Cholecystokinin and pancreozymin, one single hormone? *Acta physiol. scand., 66,* 196–202.
Kimura, T. (1937) Über die β-Hyodeoxycholsäure aus Schweinegalle. *Hoppe-Seylers Z. physiol. Chem., 248,* 280–284.
Kitts, W.D., Bailey, C.B. & Wood, A.J. (1956) The development of the digestive enzyme system of the pig during its pre-weaning phase of growth. *Can. J. Agric. Sci., 36,* 45–50.
Kowalewski, K., Secord, D.C. & Kolodej, A. (1974) Secretory function of totally isolated porcine stomach perfused ex vivo with homologous blood: effect of histamine. *Can. J. Surg., 17,* 340–347.
Kudryavtsev, A.A. (1935) Rabota slyunnykh zhelez u svinei. *Trudy vsesoyuznogo Instituta eksperimental'noi Veterinariya, 10,* 41–44.
Kudryavtsev, A.A., Kudryashov, M.V. & Troyanovskii, D.P. (1935) K kharakteristike rasstroistv pishchevaritel'nykh funktsii u porosyat. *Trudy vesesoyuznogo Instituta eksperimental'noi Veterinariya, 10,* 33–44.
Kutas, F. & Szabó, J. (1974) A study of gastric acid production in pigs in relation to early weaning. *Acta vet. Acad. Sci. hung., 24,* 133–138.
Kvasnitskii, A.V. (1951) Voprosy fiziologii pishchevareniya u svinei. *Sel'khozgiz, Moscow.*
Kvasnitskii, A.V. & Bakeeva, E.N. (1941) O zheludochnoi sekretsii i pishchevarenii u porosyat v podsosnom periode. *Trudy Inst. Svinovodstvo, Kiev, 15,* 3–42.
Laplace, J.P. (1976a) L'éxcrétion chez le porc. 1. Electromygographie des voies biliaires extra-hépatique. *Recl Méd. vét., 152,* 33–43.
Laplace, J.P. (1976b) L'éxcrétion chez le porc. 2. Electromyographie et dynamique de l'éxcrétion de bile. *Recl Méd. vét., 152,* 401–411.
Ledoux, M. & Lamy, F. (1975) Electrophoretic characterization of porcine pancreatic (pro) elastase A and B. *Can J. Biochem., 53,* 421–432.
Lehmann, G., Koppe, L., Tepper, H-J. & Hilgetag, G. (1964) Zur Gewinnung von Chenodesoxycholsäure aus Schweinegalle. *Mber. dtsch. Akad. Wiss., 6,* 274–276.
Lehmann, G., Tepper, H-J. & Hilgetag, G. (1964) Zur Isolierung von Hyocholsaure und 3 α-Hydroxy-6-oxo-5 α-cholansäure aus Schweinegallen Säuregemischen. *Biochem. Z., 340,* 75–79.
Lewis, C.J., Hartman, P.A., Liu, C.H., Baker, R.O. & Catron, D.V. (1957) Digestive enzymes of the baby pig. Pepsin and trypsin. *J. Agric. Fd. Chem., 5,* 687–693.
Lewis, U.J., Williams, D.E. & Brink, N.G. (1956) Pancreatic elastase: purification, properties and function. *J. biol. Chem., 222,* 705–720.
Louvard, M.N. & Puigeserver, A. (1974) On bovine and porcine anionic trypsinogens. *Biochim. biophys. Acta, 371,* 177–185.
Magee, D.F. & Hong, S.S. (1966) Studies on pancreatic physiology in pigs. *in* Swine in biomedical research. Eds. Busted, L.K. & McLellon, R.O. pp. 109–119. Batelle Memorial Institute, Washington.
Magee, D.F. & White, T.T. (1965) Influence of vagal stimulation on secretion of pancreatic juice in pigs. *Annls Surg., 161,* 605–607.

Marchis-Mouren, G. (1959) Contribution a l'étude de la purification de la lipase de porc et a l'étude de l'equipement enzymatique du suc pancréatique de diverses éspèces. Thesis, Marseille.
Marchis-Mouren, G. (1965) Étude comparée de l'equipement enzymatique du suc pancréatique de diverses éspèces. *Bull. Soc. Chim. biol., 47,* 2207–2217.
Marchis-Mouren, G., Charles, M., Ben Abdeljlil, A. & Desnuelle, P. (1961) Sur l'equipement en enzymes du suc pancréatique de porc et de chien. *Biochim. biophys. Acta, 50,* 186–188.
Matschiner, J.T. (1971) Naturally occurring bile acids and alcohols in the bile. Chemistry, Physiology & Metabolism. Part 1. Chemistry. pp. 11–46. Plenum Press.
Matschiner, J.T., Ratliff, R.L., Doisy, E.A. Jr., Elliott, W.H., Hsia, S.L. & Doisy, E.A. (1958) Bile acids. IX. Metabolites of hyodeoxycholic acid $24C^{14}$ in surgically jaundiced rats. *J. biol. Chem., 230,* 589–596.
Merritt, A.M. & Brooks, F.P. (1970) Basal and histamine-induced gastric acid and pepsin secretion in the conscious miniature pig. *Gastroenterology, 58,* 801–814.
Meyer, K.H., Fischer, E.H. & Bernfeld, P. (1947) Sur les enzymes amylolytiques (1). L'isolement de l'α-amylase de pancreas. *Helv. chim. Acta 30,* 64–78.
Moeller, H., Sewing, K-F, Geisbe, H. & Schmolke, M. (1974) Gastric acid secretion in conscious pigs. *Naunyn-Schmiedeberg's Arch. Pharmacol., 283,* 83–92.
Momsen, W.E. & Brockman, H.L. (1977) Purification and characterization of cholesterol esterase from porcine pancreas. *Biochim. biophys. Acta, 486,* 103–113.
Noakes, D.E. (1972) The development of digestive function in the baby pig. Ph.D. Thesis, Univ. of London.
Noakes, D.E., Cranwell, P.D. & Hill, K.J. (1968) Studies on gastric secretion in the pig. *Proc. Nutr. Soc., 27,* 2A–3A.
Northrop, J.H. (1930) Crystalline pepsin. I. Isolation and tests of purity. *J. gen. Physiol., 13,* 739–755.
Palade, G.E. (1959) Functional changes in the structure of cell components. *in* Subcellular Particles. Ed. Hayashi, T. pp64–80. Ronald, New York.
Palade, G.E., Siekevitz, P. & Caro, L.G. (1962) Structure, chemistry and function of the pancreatic exocrine cell. *in* The exocrine pancreas. pp 23–55. CIBA Foundation symposium, Churchill, London.
Peanasky, R.J., Gratecos, D., Baratti, J. & Rovery, M. (1969) Mode of activation and N-terminal sequence of subunit II of bovine procarboxypeptidase A and of porcine chymotrypsinogen C. *Biochim. biophys. Acta, 181,* 82–92.
Pekas, J.C. (1965) Permanent physiological fistula of the pancreas and other digestive glands. *J. appl. Physiol., 20,* 1082–1084.
Pekas, J.C., Thompson, A.M. & Hays, V.W. (1966) Characteristics of the exocrine pancreatic secretion of the young pig. *J. Anim. Sci., 25,* 113–121.
Phelan, J.J. & Hirs, C.H. (1970) The primary structure of porcine pancreatic ribonuclease. III. The disulphide bonds. *J. biol. Chem., 245,* 654–661.
Polivoda, D.I. & Derevins'kii, V.V. (1974) Vpliv ran'ogo vidlucennya porosyat na rist ta amilolitichny aktivnist' slinnikh zaloz. *Svinovodstvo, Kiev, 20,* 86–91.
Reboud, J.P., Marchis-Mouren, G., Cozzone, A. & Desnuelle, P. (1966) Variation in the biosynthesis rate of pancreatic amylase and chymotrypsinogen in response to a starch-rich or protein-rich diet. *Biochim. biophys. Res. Commun., 22,* 94–99.
Reinhardt, G. & Uhlenbruck, G. (1970) Elastolytic enzyme acting on neuraminic acid containing red cell glycoproteins. *Hoppe-Seyler's Z. physiol. Chem., 351,* 827–833.
Roholt, O.A. & Pressman, D. (1971) The sequence around the active-center tyrosyl of porcine carboxypeptidase B. *Eur. J. Biochem., 18,* 79–85.
Rovery, M., Charles, M., Guy, O., Guidoni, A. & Desnuelle, P. (1960) Purification de nouveaux précurseurs protéolytiques de pancréas. *Bull. Soc. Chim. biol. 42,* 1235–1246.
Scheele, G.A. & Palade, G.E. (1975) Studies on guinea-pig pancreas. Parallel discharge of exocrine enzyme activities. *J. biol. Chem., 250,* 2660–2670.
Schneeman, B.O. & Lyman, R.L. (1975) Factors involved in the intestinal feedback regulation of pancreatic enzyme secretion in the rat. *Proc. Soc. exp. Biol. Med., 148,* 897–903.
Sepulveda, P., Marciniczyn, J., Liu, D. & Tang, J. (1975) Primary structure of porcine pepsin. III. Amino sequence of a cyanogen bromide fragment CB2A and the complete structure of porcine pepsin. *J. biol. Chem., 250,* 5082–5088.

Sève, B. & Laplace, J.P. (1975) Influence de la substitution des protéines du lait par des protéines de poisson sur quelques caractéristiques du contenu gastriques chez le porcelet sevré à 12 jours. *Annls Zootech., 24,* 43–57.
Shimizu, K. (1953) The isolation of lithocholic acid from hog bile. *J. Biochem., Tokio, 40,* 69–73.
Shotton, D.M. & Hartley, B.S. (1970) Amino-acid sequence of porcine pancreatic elastase and its homologies with other serine proteinases. *Nature, Lond., 225,* 802–806.
Shotton, D.M. & Watson, H.C. (1970) Three-dimensional structure of tosyl elastase. *Nature, Lond., 225,* 811–816.
Smith, R.A. & Liener, I.E. (1967) Amino acid sequences involving the histidine residues of porcine trypsin. *J. biol. Chem., 242,* 4033–4039.
Sugiyama, G. (1937) Beitrag zur Kenntnis der 3-Oxy-6-ketoallocholansäure und Synthese der α-3, 6-Dioxy-allocholansäure. *J. Biochem., Tokio, 25,* 157–165.
Sullivan, J.F., Burch, R.E., Quigley, H.J. & Magee, D.F. (1974) Zinc deficiency and decreased pancreatic secretory response. *Am. J. Physiol., 227,* 105–108.
Swell, L., Field, H. & Treadwell, C.R. (1954) Sterol specificity of pancreatic cholesterol esterase. *Proc. Soc. exp. Biol. Med. 87,* 216–218.
Szabó, J., Ribiczeyné–Szabó, P. & Rafai, P. (1976) A pancreas-hidrolázok aktivitásának vizsgaláta különbözö környezeti hömésékleten tartott sertésekben. *Magyar Allatorvosok Lapja, 31,* 325–328.
Tartakoff, A.M., Jamieson, S.D., Scheele, G.A. & Palade, G.E. (1975) Studies on the pancreas of the guinea-pig. Parallel processing and discharge of exocrine proteins. *J. biol. Chem., 250,* 2671–2677.
Taylor, W.H. (1959) Studies on gastric proteolysis. 4. Proteinase activity of gastric juice and gastric mucosal extracts at pH 6 to 8. *Biochem. J., 71,* 626–632.
Taylor, W.H. (1962) Proteinases of the stomach in health and disease. *Physiol. Rev., 42,* 519–553.
Tschetsche, H., Wachter, E., Kupfer, S. & Niedermeier, N. (1969) Über Trypsininhibitoren IV. Die zwei spezifischen Trypsininhibitoren des Schweinepankreas. *Hoppe-Seyler's Z. physiol. Chem., 350,* 1247–1256.
Uriel, J. & Avrameas, S. (1965) Systematic fractionation of swine pancreatic hydrolases. I. Fractionation of enzymes soluble in $(NH_4)_2SO_4$ solution at 0.40 saturation. *Biochemistry, N.Y. 4,* 1740–1749.
Uram, M. & Lamy, F. (1969) Purification of two proelastases from porcine pancreas. *Biochim. biophys. Acta, 194,* 102–111.
Verger, R., de Haas, G.H., Sarda, L. & Desnuelle, P. (1968) Purification from porcine pancreas of two molecular species with lipase activity. *Biochim. biophys. Acta, 188,* 272–282.
Vodovar, N., Flanzy, J. & Francois, A.C. (1964) Intestin grêle du porc. I. Dimension en fonction de l'age et du poids, étude de la jonction du canal choledoque et du canal pancréatique a celui-ci. *Annls. Biol. anim. Biochim. Biophys., 4,* 27–34.
Voytek, P. & Giessing, E.C. (1971) Studies of an anionic trypsinogen and its active enzyme from porcine pancreas. *J. biol. Chem., 246,* 508–516.
Walker, D.M. (1959) The development of the digestive system of the young animal. I. Tissue weights, dry matter of tissues, total acidity and chloride content of stomach contents in the young pig. *J. agric. Sci., Camb., 52,* 352–363.
van Wezel, F.M. & de Haas, G.H. (1975) Phospholipase A_2 isoenzyme from porcine pancreas. Purification and some properties. *Biochim. biophys. Acta., 410,* 299–309.
Windaus, A. & Bohne, A. (1923) Über Hyo-glyko-desoxy-cholsäure und über Hyo-desoxy-cholsäure. *Justus Liebigs Annln Chem., 433,* 278–287.
Wright, R.D., Jennings, M.A., Florey, H.W. & Lium, R. (1940) The influence of nerves and drugs on secretion by the small intestine and an investigation of the enzymes of intestinal juice. *Quart. Jl. exp. Physiol., 30,* 73–120.
Yamasaki, Y., Murakami, K., Irie, M. & Ukita, T. (1968) Purification and properties of porcine pancreatic ribonuclease. *J. Biochem., Tokyo, 63,* 25–32.
Ziegler, P. (1956) The elucidation of the structure of hyocholic acid. *Can. J. Chem., 34,* 523–529

CHAPTER 6
DIGESTION OF CARBOHYDRATES

SUMMARY

The sequence of carbohydrate digestion is illustrated diagrammatically in Fig. 6.1. It begins in the buccal cavity, where the salivary α-amylase starts the breakdown of starch to a mixture of dextrins, maltose and maltotriose. This continues for a time together with some bacterial fermentation of the sugar, starch and hemicellulose in the oesophageal part of the stomach. As the contents move into the fundal region they gradually become very acid as a result of gastric secretion, so that salivary and bacterial digestion are suppressed.

As the chyme enters the small intestine, the acidity is greatly reduced by the addition of the alkaline pancreatic juice, bile and secretions of the Brunner's glands of the duodenum. This rise in pH enables the α-amylase from the pancreatic juice to continue the breakdown of starch. The dextrins and sugars from

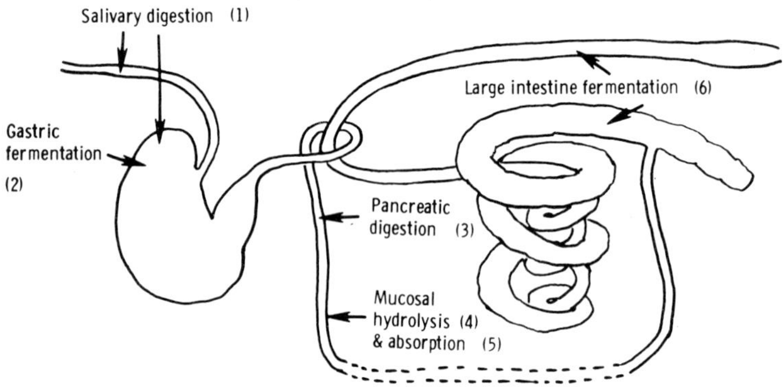

(1)	starch	→	maltose maltotriose dextrins
(2)	sugar starch hemicellulose	→	lactic acid
(3)	starch	→	maltose maltotriose dextrins
(4)	disaccharides dextrins	→	monosaccharides
(5)	monosaccharides in lumen	→	monosaccharides in blood
(6)	cellulose residual starch residual sugar	→	volatile fatty acids

Fig. 6.1 Diagram showing the sites of carbohydrate digestion

the starch, together with those present in the diet, are exposed to the action of the carbohydrases on the surface of the small intestine mucosa, and are largely split to monosaccharides. These are actively absorbed by the mucosa and pass into the capillaries leading to the portal vein.

Carbohydrates which escape these processes in the small intestine pass into the large intestines where they are exposed to a large population of micro-organisms which ferment the soluble sugars, the starch and some of the cellulose, with the production of volatile fatty acids, which are absorbed through the large intestine mucosa, and so pass into the portal blood.

SALIVARY DIGESTION

As described in Chapter 5, saliva contains significant quantities of only one enzyme, α-amylase. Detailed studies of the action of salivary amylase have been made almost entirely on human saliva but the pig enzyme appears to have precisely similar properties. Like pancreatic amylase, it breaks starch down to a mixture of maltose, maltotriose and various dextrins, and it requires the presence of chloride for full activity (Myrbäck, 1926). It is active over the pH range 3.8–9.4 with an optimum, in the presence of chloride, of 6.9 (Bernfeld, Staub & Fischer, 1948).

Whelan & Roberts (1953) showed that with amylose, a linear starch molecule where the glucose units are joined only by α 1–4 links, the hydrolysis proceeded as if all these links except the terminal one at each end were equally susceptible so that a mixture of maltose and maltotriose in the ratio of 2.25:1 resulted. Walker & Whelan (1960) showed that, in the presence of high concentrations of salivary amylase, maltotriose was very slowly hydrolysed to maltose and glucose. Thus the terminal links are not totally resistant to salivary α-amylase.

With branched chain starches such as amylopectin and glycogen which consist of α 1–4 glucose chains joined to one another by α 1–6 branch links, the branch link and some links next to it are resistant to α-amylase hydrolysis (Roberts & Whelan, 1960; Fig. 6.2). The result is that, in addition to the production of maltose and maltotriose from the parts of the chain remote from a branch point, a variety of larger oligosaccharides, called "α-limit dextrins", are produced which may incorporate one or more 1–6 branch links (Roberts & Whelan, 1960; Bines & Whelan, 1960; Walker & Whelan, 1960). Fig. 6.3 (Roberts & Whelan, 1960) illustrates a number of these limit dextrins whose structure was determined.

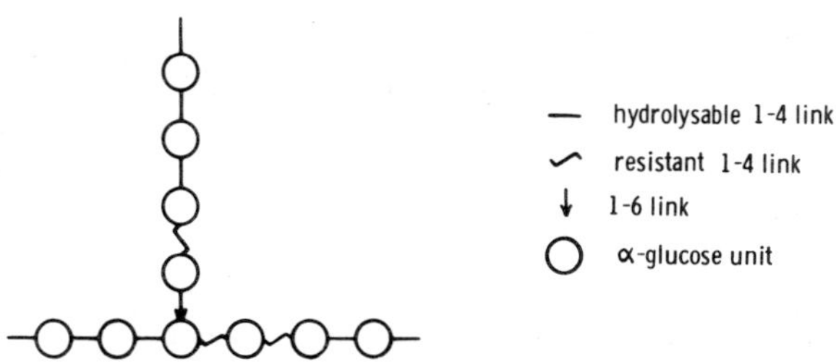

Fig. 6.2 Linkages in starch or glycogen resistant to α-amylase. (from Roberts & Whelan, 1960)

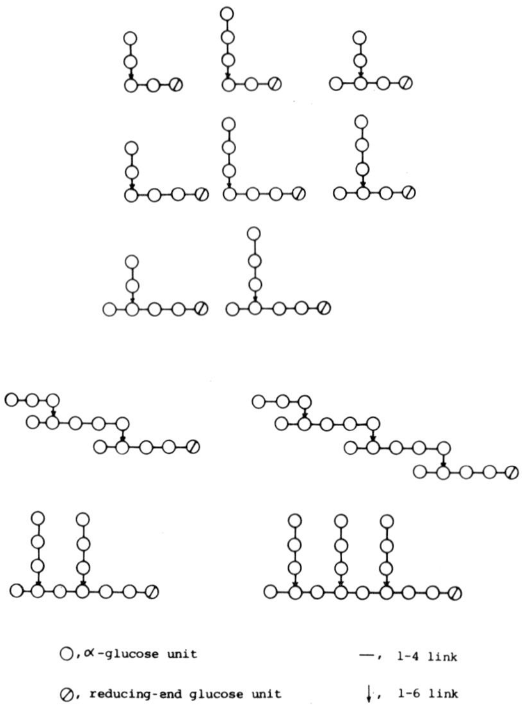

Fig. 6.3 *α-Limit dextrins resulting from salivary digestion. (from Roberts & Whelan, 1960)*

GASTRIC DIGESTION OF CARBOHYDRATES

Salivary

Salivary action, which has started in the buccal cavity, continues in the oesophagus and the oesophageal part of the stomach. As the food is mixed with the gastric secretions, particularly in the fundus of the stomach, the pH falls, and below pH 3.6 the amylase action stops. Arkhipovets (1962) inserted starch-filled tubes through fistulas into the stomachs of pigs and showed that starch digestion there was quite appreciable, especially with dry cereal diets which stimulated salivary secretion.

Fermentation

In the pig, bacterial fermentation of carbohydrate occurs both in the stomach and in the large intestine, some sugar and starch and a little hemicellulose being fermented in the stomach, and residual sugar and starch, together with cellulose, in the large intestine. This bacterial digestion in different parts of the tract is described in detail later in this chapter.

DIGESTION OF CARBOHYDRATES IN THE SMALL INTESTINE

On passage along the duodenum, the acidity of the chyme from the stomach is almost completely neutralized by the alkaline pancreatic juice and, to a lesser extent, the bile and the secretions of the Brunner's glands in the duodenal mucosa. The pH rises gradually to a level which is suitable for the action of pancreatic amylase and the mucosal carbohydrases.

Pancreatic digestion

The pancreatic juice of the pig, like that of other mammals, contains a high concentration of an α-amylase which continues the breakdown of starch. Pancreatic amylase has a very similar action to that already described for salivary amylase, converting starch to maltose, maltotriose and α-limit dextrins together with traces of glucose (Meyer & Gonon, 1951; Caldwell, Adams, Kung & Toralballa, 1952). Pig pancreatic α-amylase differs in one respect from salivary α-amylase in that, under optimum conditions for its action, it does not act equally on all susceptible bonds (Robyt & French, 1967, 1970a, b) and, in the initial stages of hydrolysis, a larger amount of reducing sugar, particularly maltose, is produced, in proportion to the reduction in chain length, than would be the case with random hydrolysis (Robyt & French, 1967; Banks, Greenwood & Khan, 1970).

In addition to its presence in the lumen of the gut, pancreatic amylase is also adsorbed on to the mucosal surface of the small intestine. This phenomenon, originally described in the rat (Ugolev, 1960; Dahlqvist & Thomson, 1963), has been shown in the pig (Ruttloff, Friese & Täufel, 1967a, b; Stevens & Kidder, 1972), but its importance is not known.

Mucosal digestion

Hydrolysis

The mucosal cells of the small intestine produce a number of carbohydrases which are located on the brush border of the mature cells, and so are in contact with the intestinal lumen. This set of enzymes, first reported and studied in the pig, and subsequently found in other non-ruminant animals, consists of lactase, trehalase and four maltases.

LACTASE

A capacity to hydrolyse lactose was first demonstrated in pig small intestine mucosa by Mendel & Mitchell (1907), and is due to an enzyme on the mucosal brush border. This enzyme will hydrolyse a number of β-galactosides and will also, to some extent, transfer β-galactosyl groups to another sugar (Wallenfels & Bernt, 1953). Other enzymes with β-galactosidase activity, but no direct digestive function, also occur within small intestine mucosal cells. This multiplicity of β-galactosidases, originally demonstrated in other mammals, occurs in the pig (Sato & Yamashina, 1974). The digestive lactose-splitting enzyme is usually known as "neutral lactase" or "brush border lactase."

Dahlqvist (1962a) obtained a partially purified preparation of brush border lactase of pig small intestine mucosa and found it to have an optimum pH of 6.0 and relative activities towards lactose, cellobiose and gentiobiose of 100,

25–30 and 17 respectively. Heyworth & Dahlqvist (1962) confirmed that these activities were all due to the same enzyme which is thus both a β-galactosidase and a β-glucosidase. The main physiological substrate of intestinal brush border lactase is lactose, the predominant sugar of most mammalian milks including that of the sow. Lactase also acts on other, chemically similar sugars which occur in milk, usually, as in the pig, only as trace components. As might be expected, this enzyme is present at very high levels in the small intestine mucosa of very young mammals and its concentration decreases with age (Plimmer, 1907). The decrease in lactase level with age has been studied particularly extensively in the pig by a number of workers whose results all show the same general trends. Due to differences in method, quantitative comparisons between the results of different authors are very rarely possible. Some workers homogenized whole gut and others only mucosa, and nearly all used different assay methods giving different units which were then reported in some cases per g mucosa or tissue and, in others, per g protein. Only the later workers inhibited the internal cellular lactases, so the earlier measurements will have included a contribution from these non-digestive enzymes. The size of this contribution would depend on the pH chosen for the assay, the age of the animal and the site along the intestine which was sampled. Nevertheless, the results of the later, more detailed, and more specific studies agree, in general, with those of the earlier workers. These were summarized by Manners & Stevens (1972), who, in detailed studies of change in distribution of total lactase in small intestine mucosa with age, showed that the fall in level was rapid only for the first week (Fig. 6.4). After this, there was little change other than a gradual disappearance of lactase from the distal end (Fig. 6.5). Ekstrom, Benevenga & Grummer (1975a) confirmed the fall in specific activity with age and the more marked decline with age in the distal half of the small intestine (Table 6.1). Widdowson, Colombo & Artavanis (1976) found that the level of neutral lactase in the mucosa of the newborn pig, expressed per gram of protein, fell by about a third on feeding for a day, while the total quantity of lactase in the gut rose. In 10 days, the lactase activity per gram of protein had fallen to below half the birth level. At both ages the level in the proximal small intestine was only slightly higher than in the distal. This is in agreement with the findings of other workers that the fall in lactase level at the distal end comes later. Kidder & Manners (unpublished) looking at the brush border lactase, mainly in older pigs, showed that the decline in lactase level with age became more noticeable again in pigs of over 2 years of age (Fig. 6.6). This pattern of a rapid fall in lactase in the very young animal, and a further fall at a much greater age, is in keeping with the findings in other mammals.

Ekstrom, Benevenga & Grummer (1975a) showed that mucosal lactase levels at 21 weeks of age were not significantly affected by whey feeding in place of the conventional weaning diet. In other words, the presence of the substrate did not influence the decline in enzyme level. However, they found that lactase levels were significantly higher in their Chester White pigs than in their Hampshires (Ekstrom, Benevenga & Grummer 1975a, b; Ekstrom, Grummer & Benevenga, 1976; Fig. 6.7). Breed differences have not been reported up to the present by other workers.

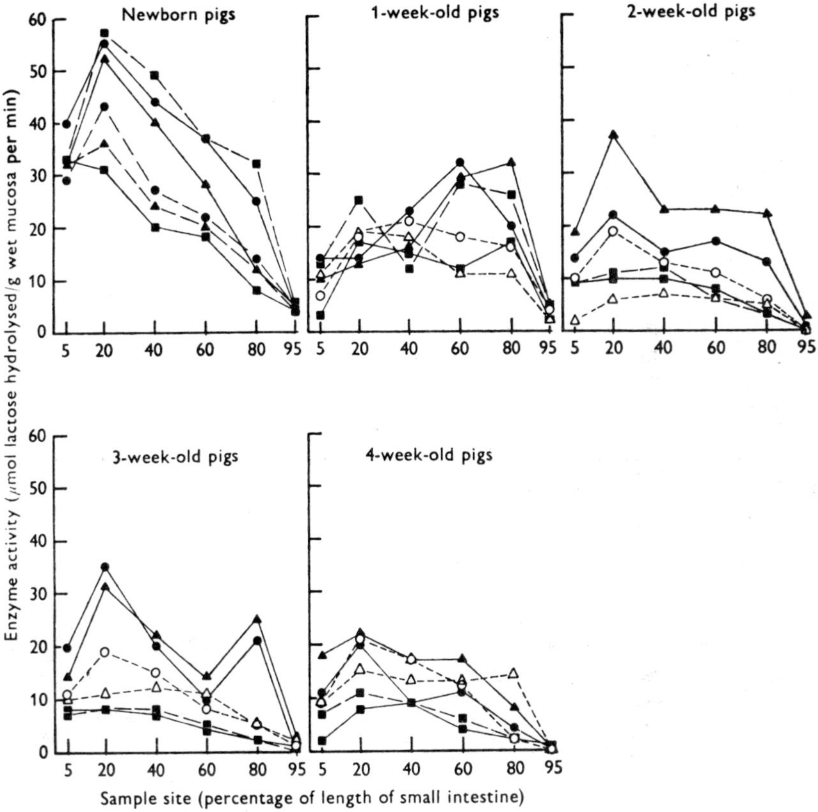

Litter 1, sow-reared, ■——■, ■— —■. Litter 2, sow-reared, ▲——▲, ▲— —▲; artificially reared, △ - - - △. Litter 3, sow-reared, ●——●, ●— —●; artificially-reared, ○ - - - ○.

Fig. 6.4 Lactase activity in the mucosa at six standardized sites along the small intestine of individual members of pairs of pigs from three litters of Large White X Wessex pigs killed at weekly intervals from birth to 4 weeks of age. (from Manners & Stevens, 1972)

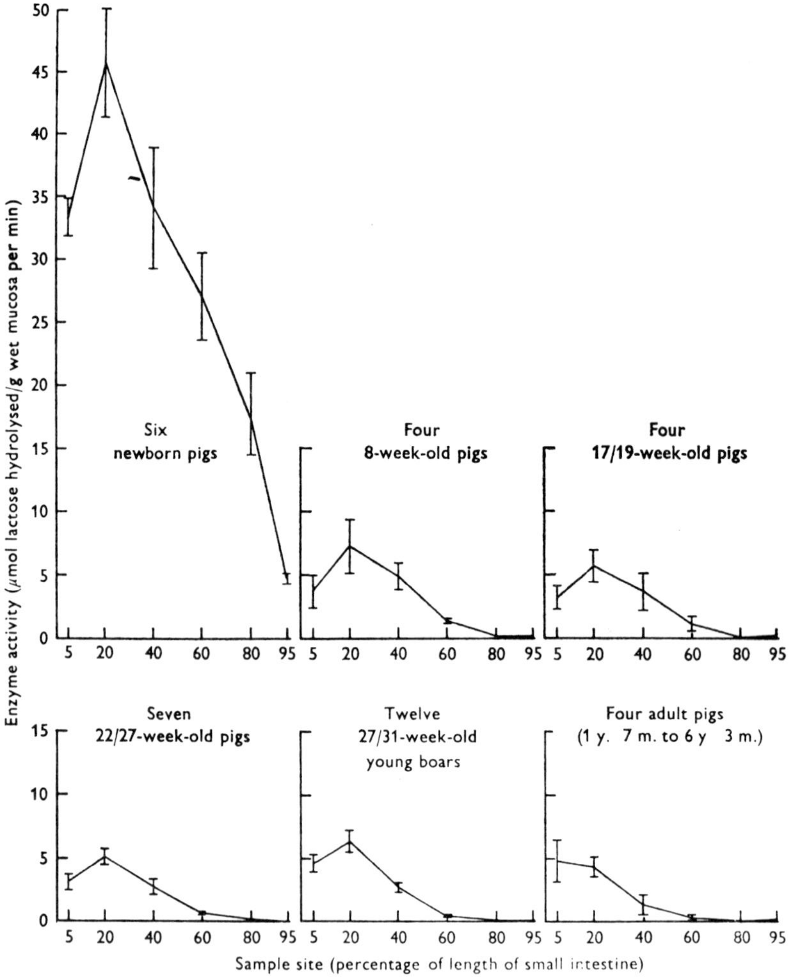

Fig. 6.5 *Lactase activity at six standardized sites along the small intestine of pigs killed between birth and maturity. The vertical bars represent the standard errors of the mean. (from Manners & Stevens, 1972)*

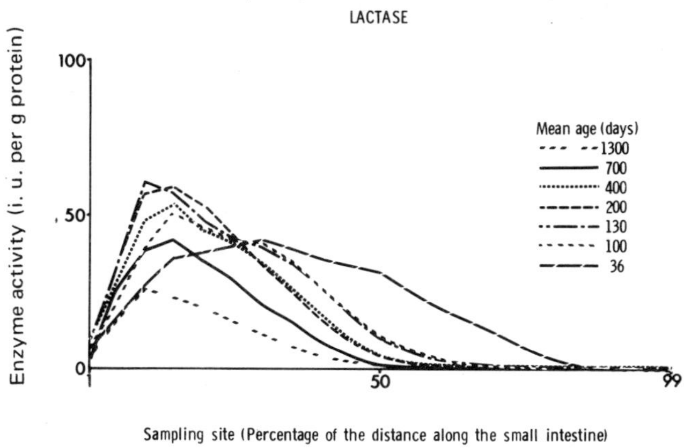

Fig. 6.6 Changes with advancing age in lactase activity along the small intestine of pigs (seven animals in each age group). (Kidder & Manners, unpublished)

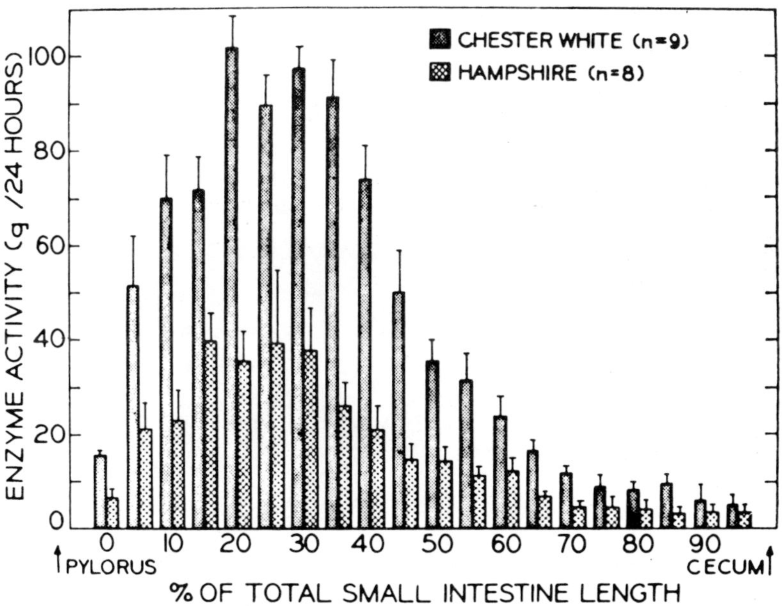

Fig. 6.7 Mucosal lactase activity along the small intestine of Chester White and Hampshire pigs. The height of the columns represents the total lactase activity in the 1 metre length of intestine. (from Ekstrom, Benevenga & Grummer, 1975a)

Table 6.1 *Changes in lactase activity with age in the pig (Ekstrom, Benevenga & Grummer, 1975a)*

Total lactase activity in the distal 50% of the small intestine at various ages

Age	Chester White			Hampshire		
	No.	grams 24 hr	% of total activity	No.	grams 24 hr	% of total activity
days						
1	4	60±44	32.1	4	64±44	36.4
8	4	226±44	45.7	4	258±44	54.8
15	4	351±44	47.9	4	248±44	50.1
22	4	300±44	41.8	4	164±44	45.7
29	4	189±44	34.8	4	96±44	26.7
43	4	100±44	24.7	4	80±44	24.8
100–110	4	106±24.4	21.4	7	74±10.4	23.3
150–180	33	92± 5.9	13.4	18	51± 6.9	19.5

TREHALASE

Trehalase was isolated from pig small intestine mucosa by Dahlqvist (1960a) who regarded it as specific for α–α' trehalose, and found a pH optimum of 6.0. The specificity of the enzyme was confirmed by Labat, Baumann & Courtois (1973) who showed that it broke the trehalose down to two α-glucose molecules. Trehalose occurs in most insects, many fungi and a few plants, and is unlikely to be a significant item in the diet of the modern commercial pig. Dahlqvist (1961b) and Stevens & Kidder (1972) showed that it had a predominantly proximal distribution along the small intestine. Dahlqvist (1961c) showed it to be absent from the mucosa of the new-born pig, and Kidder & Manners (1976) found that its level increased with age up to about 200 days, the distribution being similar and predominantly proximal at every age studied (Fig. 6.8). The proximal distribution is consistent with the relatively low optimum pH.

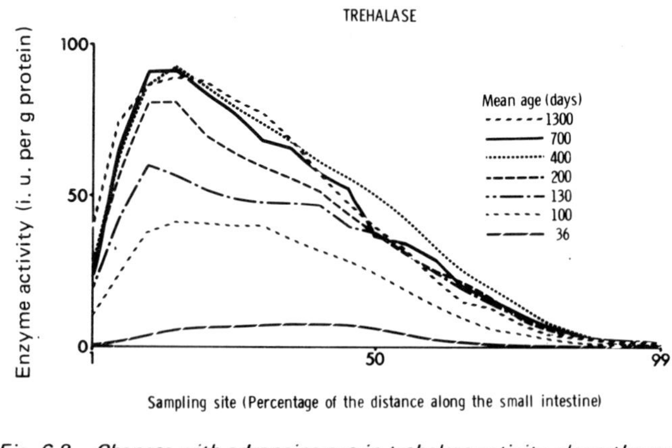

Fig. 6.8 *Changes with advancing age in trehalase activity along the small intestine of pigs (seven animals in each age group). (Kidder & Manners, unpublished)*

MALTASES

Brown & Heron (1880) showed maltase and sucrase activity in pig small intestine mucosa, and Larner & McNickle (1955) and Larner & Gillespie (1956) showed that an enzyme also occurred which would hydrolyse the 1–6 glucosidic bonds of isomaltose. Dahlqvist (1958, 1959a, 1960b, c, d, e, 1961d) showed that these activities in pig small intestine mucosa were due to a group of four enzymes, all of which rapidly hydrolysed maltose, and which he therefore called maltases, but which also had other enzymic actions, some of which occurred in only one of these maltases, and some actions which occurred in several. A similar pattern of four maltases has been found in other non-ruminant mammals (Semenza, 1968). Many carbohydrase activities of small intestine mucosa have been shown to be due to one or more of these maltases, and Table 6.2 summarises the main enzymic actions of the four maltases using the

Table 6.2 Activities of mucosal maltases (classified according to Dahlqvist, 1962)

	Other names	Chief substrates other than maltose
Ia	isomaltase	isomaltose, limit dextrins
Ib	sucrase	sucrose
II	glucoamylase	maltodextrins, starch, isomaltose, limit dextrins
III	glucoamylase (heat stable maltase)	maltodextrins, starch, isomaltose, limit dextrins

classification of Dahlqvist (1962).

While each of the maltases acts on a number of substances, they are completely specific for the α-glucosyl portion of the substrate (Dahlqvist 1958, 1959b, 1960b, f) and the reaction results in the production of α-glucose (Semenza, Curtius, Kolinská & Müller, 1967), or, to some extent with concentrated substrates, the transfer of the α-glucosyl group with the formation of a new α-glucoside (Dahlqvist, 1959b, 1960b). Maltase Ia, because of its particularly rapid action on isomaltose, is usually referred to as isomaltase while maltase Ib, the only one which acts on sucrose, is usually referred to as sucrase. These two enzymes have been shown in the rabbit (Kolinská & Semenza, 1967) and in man and the rat, to be present as a double molecule known as the sucrase — isomaltase complex, with an apparent molecular weight of about 280,000 which can be split by chemical procedures to give the two separate enzymes (Conklin, Yamashiro & Gray; 1975). Studies on human patients with a genetic deficiency of this complex make it clear that this double enzyme is synthesised as a unit (Auricchio, Ciccimarra, Moauro, Rey, Jos & Rey 1972; Preiser, Menard, Crane & Cerda, 1974). Sucrase and isomaltase levels in pig mucosa are closely correlated (Kidder & Manners, 1976), indicating the likelihood of a similar complex in the pig.

In the pig, both maltase Ia and Ib have a fairly narrow pH optimum between 6 and 6.5 (Dahlqvist, 1960c) while maltases II and III have broader optima between 6.5 and 7.5 (Dahlqvist 1960b).

Hydrolysis of sugars by maltases

Sucrose, a major constituent of molasses, and present in much plant material, is only hydrolysed by maltase Ib, but this enzyme has very high activity against sucrose and is present at a high concentration in the mucosa of all but very young pigs.

A number of less common sugars are also α-glucosides and are hydrolysed by pig mucosal maltases, some like maltosucrose (Dahlqvist, 1960f) or leucrose (Ruttloff, Friese & Täufel, 1964) very rapidly, others, like turanose (Dahlqvist, 1960f) or palatinose (Dahlqvist, 1961d), at a moderate speed, and some, like melezitose, very slowly (Dahlqvist, 1960f). Dahlqvist (1960e) gave the relative rates at which a number of sugars were hydrolysed by a pig mucosal homogenate and by the separate maltases obtained from it. He summarized this information and gave the relative contributions of each enzyme where more than one acted on the same substrate (Dahlqvist, 1960f; Tables 6.3 & 6.4). As the existence of the sucrase-isomaltase complex was not understood at this time, it is uncertain whether the maltase I and the sucrase referred to in these tables is the complex, or sucrase alone. Palatinose is only hydrolysed by enzymes with isomaltase action and is sometimes used to measure the level of this activity (Burgess, Levin, Mahalanabis & Tonge, 1964).

Table 6.3 *Substrates for the three different hog intestinal maltases. Relative rate of hydrolysis measured at 0.1388 M substrate concentration. (The rate of hydrolysis of maltose has been arbitrarily set at 100) (Dahlqvist, 1960f)*

Enzyme	Substrates (relative rate of hydrolysis within parenthesis)
Maltase I (invertase)	Maltose (100) Sucrose (170) Turanose (7.5) Maltosucrose (80)
Maltase II	Maltose (100) Isomaltose (2) Turanose (4) Phenyl-α-D-glucopyranoside (3) Maltosucrose (40) Melezitose (0.1)
Maltase III	Maltose (100) Isomaltose (5) Turanose (6) Phenyl-α-D-glucopyranoside (5) Maltosucrose (40) Melezitose (0.4)

Table 6.4 Relative rate of hydrolysis of different α-D-glucopyranosides by crude hog intestinal mucosa homogenates. The enzymes responsible for the hydrolysis of each substrate are listed. For each enzyme the fraction of the total activity on that substrate is given within parenthesis. (Dahlqvist, 1960f). (The figure for trehalose hydrolysis applies to the upper jejunum only)

Substrate	Relative rate of hydrolysis by crude preparation	Enzyme	
Maltose	100	Invertase (maltase I) Maltase II Maltase III	(30–35% each)
Sucrose	50	Invertase (maltase I)	(100%)
Trehalose	25	Trehalase (100%)	
Isomaltose	8–10	Specific isomaltase (70–75%) Maltase II (5–10%) Maltase III (15–20%)	
Turanose	7–8	Inverase (maltase I) (50–55%) Maltase II (15–20%) Maltase III (25–30%)	
Phenyl-α-D-glucopyranoside	5–6	Maltase II (15–20%) Maltase II (30–35%) Specific isomaltase? (50%)	

Hydrolysis of starch by maltases

The most important substrates of the maltases are the products of hydrolysis of starch. The end products of starch digestion by salivary and pancreatic α-amylases are maltose, maltotriose and α-limit dextrins, described under salivary digestion. Maltose is hydrolysed rapidly by all the maltases, and so is maltotriose, at least in man (Messer & Kerry, 1967). The 1–6 links of the limit dextrins are rapidly hydrolysed by the maltase Ia (isomaltase), and this appears to be the most important function of this enzyme. Maltase II and III also slowly hydrolyse the 1–6 links. The resulting portions of the limit dextrins may then, depending on their size, be acted on by α-amylase or maltases or both.

Partly hydrolysed starch, even before breakdown to dextrins, can be hydrolysed with the production of glucose by enzymes from the brush border of the pig small intestine mucosa (Ruttloff, Friese & Täufel, 1967b, 1968). These workers showed that, while brush border enzymes hydrolysed soluble starch they did so much less rapidly than the pancreatic α-amylase present in a mucosal homogenate. On the other hand, they showed that maltodextrins of 5 glucose units or less were more rapidly acted on by the brush border enzymes than by pancreatic α-amylase. This glucoamylase (E.C. 3.2.1.3.) hydrolysis of starch is due to the α-glucosidase action of the maltases which release glucose residues, one at a time, from the non-reducing chain end. Chromatographic studies on human jejunal mucosa reported by Auricchio & Ciccimarra (1970) and Auricchio, Ciccimarra, Starace, Vegnente, Giliberti & Provenzale (1971) showed that a little of this activity was due to the sucrase-isomaltase complex, which was particularly active against the lower molecular weight fractions of the soluble starch (D.P. approximately 30). The bulk of their glucoamylase activity, especially that against the higher molecular weight starches (D.P. approximately 575) was due to maltases II and III. Kidder, Hill & Stevens (1972), in heat inactivation studies on pig small intestine mucosal homogenates, obtained results which would have indicated that two thirds of the activity was from maltase II, one third from maltase III, and very little from the sucrase-isomaltase complex. As glucoamylases only act at the chain end, they have little detectable action on unhydrolysed starch which has long chains, and few chain ends. This has been shown in mucosal homogenates of man (McMichael & Dahlqvist, 1968; Alpers & Solin, 1970) and the pig (Kidder, Hill & Stevens, 1972). Thus the pancreatic α-amylase and the mucosal glucoamylase have complementary actions, the pancreatic enzyme acting rapidly on the long and medium chain starches and less rapidly on the very small dextrins, while the glucoamylases act on the medium, and, especially, on the small products of starch hydrolysis, releasing glucose, which can be absorbed rapidly by the mucosa.

Hydrolysis of dextran

Dextran is a bacterial polysaccharide containing glucose units joined mainly by α 1–6 glucoside links. The hydrolysis of this substance by small intestine mucosa is due to the maltases with isomaltase activity (Dahlqvist, 1963). Dextran, like palatinose, can therefore be used as a substrate to measure the α 1–6 glucosidase activity of the mucosa.

DEVELOPMENT AND DISTRIBUTION OF THE MALTASES

Total maltase

The maltase activity of pig small intestine was found by Bailey, Kitts & Wood (1956) to be very low at birth and to increase rapidly in the first few weeks of life. This was confirmed by Walker (1959), Hartman, Hays, Baker, Neagle & Catron (1961) and Tacu & Bianu (1974). All of these authors reported that the mucosal maltase level rose only for the first few weeks, after which it remained stationary, or even decreased slightly. This finding is in conflict with more recent studies on specific maltases (Manners & Stevens, 1972; Kidder & Manners, 1976) where it was found that the rise in level continued into adult life. This difference in findings could be due to the limited number of older pigs used by the earlier workers, the individual variation always encountered, and the loss of enzyme from the brush border which occurs particularly with older pigs if the sample is not taken sufficiently rapidly after slaughter.

Sucrase

Apart from the finding by Dahlqvist (1961c) that isomaltase was absent from the small intestine mucosa of one newborn pig which he examined, sucrase was, until recently, the only individual maltase which had been examined from the point of view of development with age. It had been assumed on the basis of the very poor performance of very young piglets on sucrose diets that sucrase was lacking in their digestive system (Becker, Ullrey & Terrill, 1954). The truth of this assumption was demonstrated by Bailey, Kitts & Wood (1956) who found that the sucrase level of pig small intestine, like the total maltase, was negligible at birth but rose rapidly for the first few weeks of life. This finding was confirmed by Walker (1959), and in general by Hartman, Hays, Baker, Neagle & Catron (1961), although these workers did find appreciable levels at birth.

Manners & Stevens (1972), measuring sucrase at 6 positions along the small intestine, also found very low levels at birth increasing rapidly in the first few weeks of life, and then more slowly, but with occasional pigs in each age group having much higher levels than the others (Fig. 6.9). Kidder & Manners (1976) measuring sucrase at 21 positions along the small intestines of older pigs found that the levels continued to rise up to at least 2 years of age (Fig. 6.10).

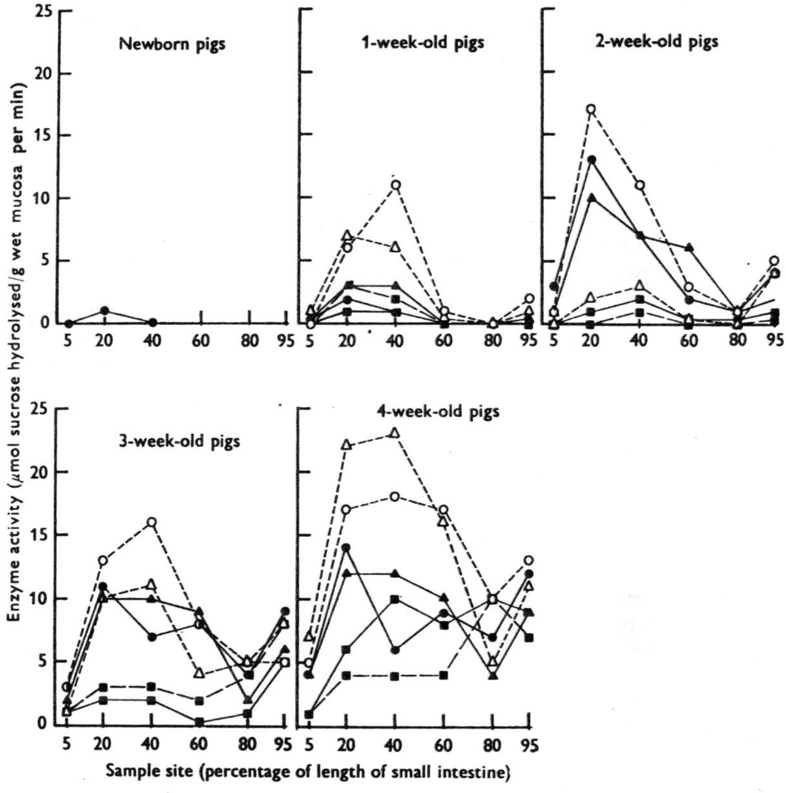

Litter 1, sow-reared, ■——■, ■— —■. Litter 2, sow-reared, ▲——▲, ▲— —▲; artificially-reared, △---△. Litter 3, sow-reared, ●——●, ●— —●; artificially-reared, ○---○.

Fig. 6.9 Sucrase activity in the mucosa at six standardized sites along the small intestine of individual members of pairs of pigs from three litters of Large White X Wessex pigs killed at weekly intervals from birth to 4 weeks of age. (from Manners & Stevens, 1972)

Fig. 6.10 Changes with advancing age in sucrase activity along the small intestine of pigs (seven animals in each age group). (Kidder & Manners, unpublished)

Isomaltase

Kidder & Manners (1976) found that the dextranase levels also rose with age throughout the period studied (Fig. 6.11). The dextranase activity is mainly due to maltase Ia (isomaltase), but there is also a slight contribution from maltases II and III. The individual dextranase results showed close correlation with the corresponding sucrase values, in keeping with the assumption that most of the dextranase activity is due to the sucrase-isomaltase complex.

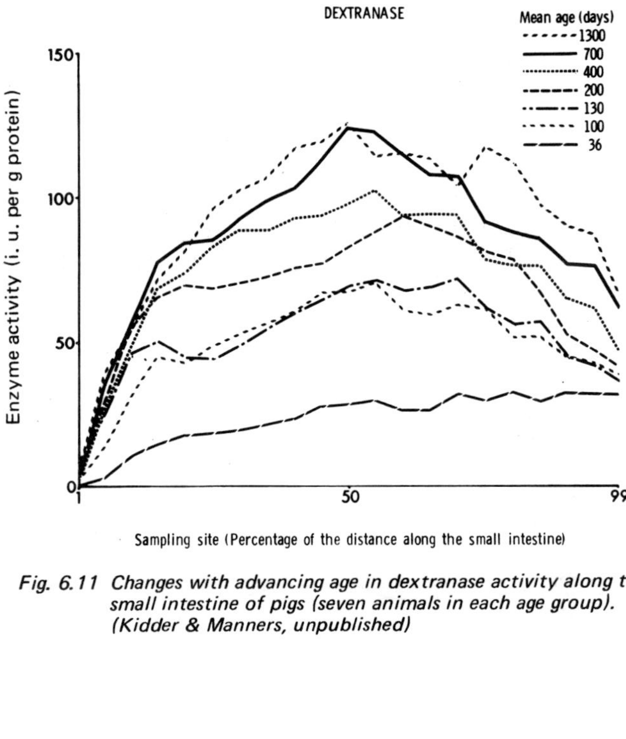

Fig. 6.11 Changes with advancing age in dextranase activity along the small intestine of pigs (seven animals in each age group). (Kidder & Manners, unpublished)

Maltases II & III

Kidder & Manners (1976) found that the levels of maltase II and III also rose with age, but this rise did not appear to continue beyond about 200 days of age (Figs. 6.12 & 6.13).

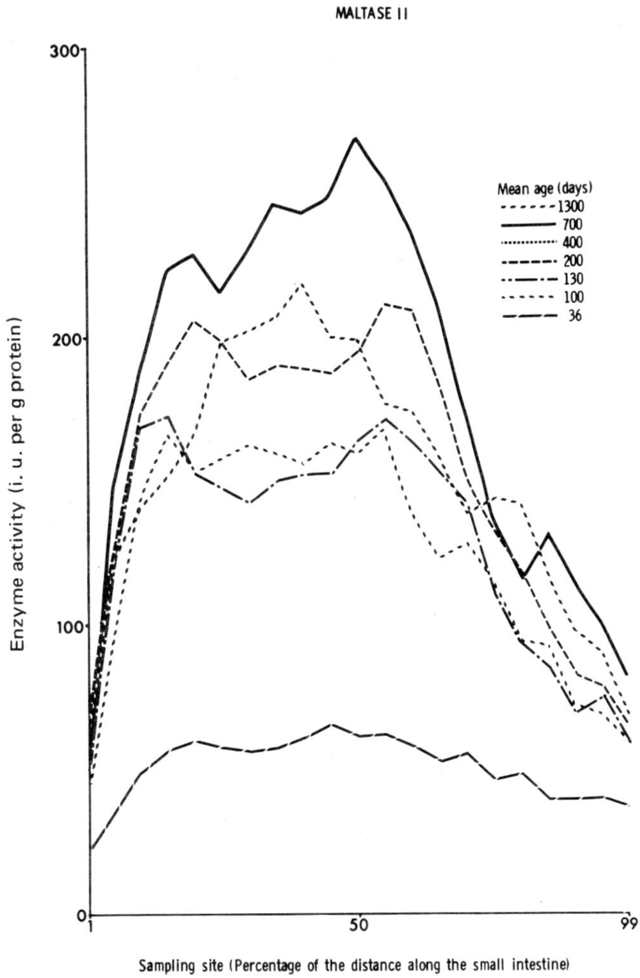

Fig. 6.12 Changes with advancing age in maltase II along the small intestine of pigs (seven animals in each age group). (Kidder & Manners, unpublished)

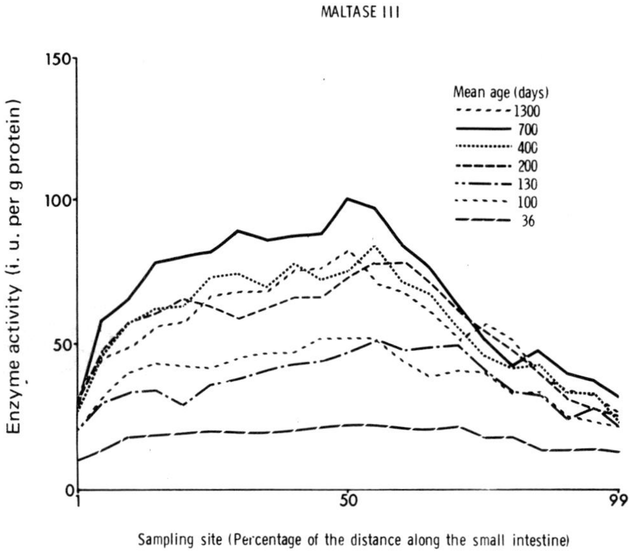

Fig. 6.13 Changes with advancing age in maltase III activity along the small intestine of pigs (seven animals in each age group). (Kidder & Manners, unpublished)

Fluctuations in enzyme level along the intestine

As described in Chapter 2, there is, in the pig, a clear-cut distinction between the regions of the small intestine containing food and those which are empty. Stevens & Kidder (1972), when measuring mucosal enzyme levels, kept a record of which regions of the small intestine were dilated with contents at the time of dissection, and which regions were closed and almost free of food material. In those regions which were dilated, they reported the sucrase and glucoamylase levels to be lower than expected, and the levels of adsorbed α-amylase to be unusually high, while in the regions without contents, the sucrase and glucoamylase levels were high, and adsorbed α-amylase almost absent. Lactase and trehalase levels seemed to be little affected by these differences in the state of the intestine.

In a larger study, Kidder & Manners (unpublished) found that this depression of enzyme activity in the mucosa of some regions of the small intestine involved all four maltases, but especially maltase II, while trehalase and lactase were usually unaffected. Adsorbed α-amylase was always present, frequently at a high level, in these regions, which nearly always corresponded with the presence of digesta in the lumen of the gut at the time of its removal from the pig. Figs. 6.14 and 6.15 show the enzyme levels along the small intestines of two pigs and the degree of dilation of the intestine recorded on dissection.

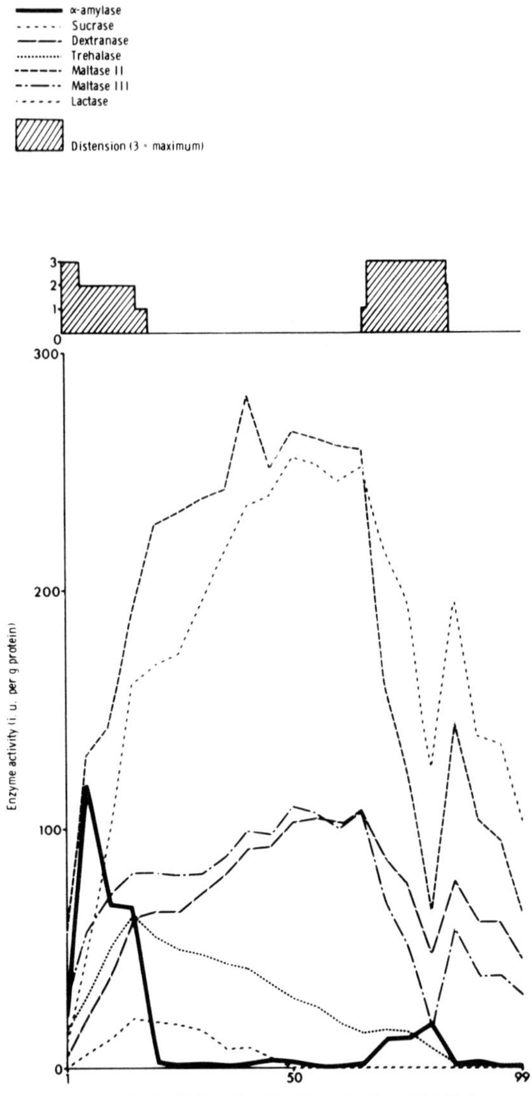

Fig. 6.14 Locations of distended regions and levels of α-amylase and the six mucosal carbohydrases along the small intestine of a sow aged 666 days. (Kidder & Manners, unpublished)

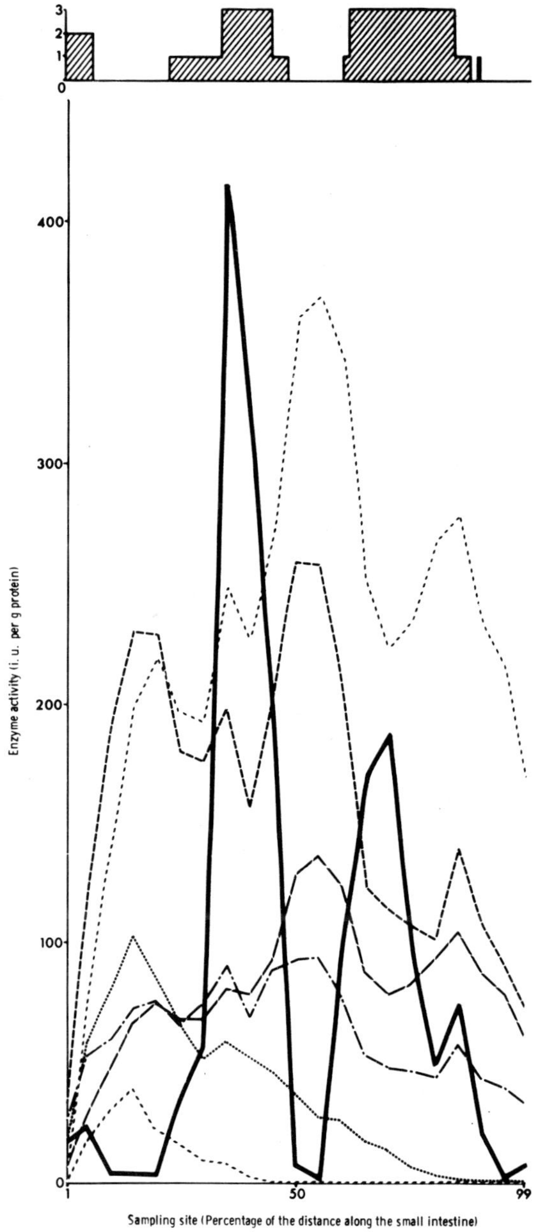

Fig. 6.15 Locations of distended regions and levels of α-amylase and the six mucosal carbohydrases along the small intestine of a boar aged 734 days. (See Fig. 6.14 for key) (Kidder & Manners, unpublished)

Several studies in other species may help to explain these regions of altered enzyme levels in the pig. Loginov (1973) found that diversion of pancreatic juice from the small intestine in rats resulted in a rise in mucosal sucrase levels. Arvanitakis & Olsen (1974), in human patients with pancreatic insufficiency, similarly found a rise in mucosal sucrase. Alpers & Tedesco (1975) found that, in rats, after 95% pancreatectomy, there were rises in mucosal sucrase and maltase, but not in trehalase and alkaline phosphatase. They showed that this change was due to release of the maltases from the brush border into the lumen by the action of pancreatic proteases, particularly elastase, an action which occurred both in vivo and in vitro. Release of enzymes from the mucosal brush border membrane by the action of elastase has also been shown by Maestracci (1976).

The presence of food material with high concentrations of pancreatic enzymes in a particular zone of the small intestine in the period immediately before and after slaughter, could therefore be expected to lead to a reduced level of sucrase and the other maltases, while the level of trehalase is unaffected. The high level of adsorbed α-amylase in these regions is also to be expected in view of the readiness with which α-amylase is known to be adsorbed by the mucosa, and subsequently gradually removed (Ugolev, 1960; Iezuitova, de Laey & Ugolev, 1964; Alpers & Solin, 1970).

Absorption

ABSORPTIVE PROCESSES

The processes of sugar absorption by the small intestine mucosa have not been studied in the pig, but they appear to be similar in a wide range of species, and observations on the relative rates of absorption of different sugars in the pig indicate that the pig resembles other species in intestinal sugar absorption. The absorption processes have been reviewed by Crane (1960, 1968), Wilson (1962) and Gray (1970) among others. It is appropriate here to summarize the main findings.

While traces of disaccharides, especially those not hydrolysed by the mucosa, are absorbed from the gut lumen (Täufel, Krause, Ruttloff & Maune, 1967), the bulk of the dietary carbohydrate is absorbed as monosaccharide. Different monosaccharides are absorbed from the lumen of the gut at different rates, glucose and galactose being absorbed rapidly, fructose much more slowly, and all other sugars at a still slower rate (Cori, 1925). This is because they pass through partly by diffusion but mainly by various absorption processes which are specific for individual sugars or groups of sugars. The most important of these processes brings about the very rapid and efficient absorption of glucose and galactose with the simultaneous transport of Na^+ from the luminal to the serosal side of the gut wall (Crane, 1960). It also assists to a slight extent the transport of some other natural sugars, D- and L-xylose, D- and L-arabinose, D- and L-fucose, L-rhamnose, D-mannose and myo-inositol being listed by Barnett, Ralph & Munday (1970). Ferguson & Burton (1977) have shown that reconstituted phospholipid vesicles containing brush border protein from mucosal cells of piglet small intestine would selectively transport glucose, and that this transport was influenced by sodium ions or phlorizin in the same way as that in intact mucosa in species in which it has been studied. A separate mechanism has been shown to exist for fructose (Gracey, Burke & Oshin, 1972), in addition to which, fructose is partly converted to glucose within the mucosa in some species (Ginsburg & Hers, 1960).

Sugars not affected by either of these specific processes, and so passing through the wall only by simple diffusion, are slowly and inefficiently absorbed.

There is some evidence that some disaccharidases, particularly sucrase, also assist in the absorption of monosaccharides from disaccharides in the lumen. The constituent monosaccharides are transported through the intestinal wall more rapidly when presented to the mucosal side as sucrose, maltose, isomaltose or trehalose than when presented as monosaccharides (Ramaswamy, Malathi, Caspary & Crane, 1974).

FERMENTATION

Microbial fermentation in the alimentary tract, which plays a significant part in the digestion of carbohydrates in the pig, has been reviewed in detail by Cranwell (1968). Barcroft, McAnally & Phillipson (1944) demonstrated that the venous blood leaving the alimentary tract of the pig, particularly the caecum and colon, contained higher levels of volatile fatty acids than the peripheral blood. The origin of these organic acids was established in studies on the gut contents of slaughtered pigs. Elsden, Hitchcock, Marshall & Phillipson (1946) demonstrated

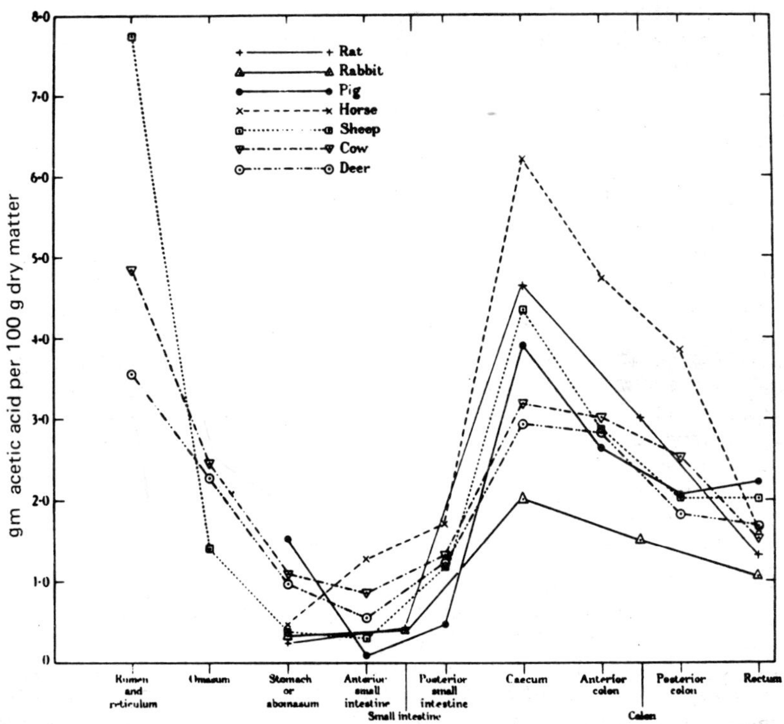

Fig. 6.16 Average concentrations of volatile fatty acids in the contents of various sections of the alimentary tract of different animals. (from Elsden, Hitchcock, Marshall & Phillipson, 1946)

the presence of volatile fatty acids (VFA) in the contents of the stomach and all parts of the large intestine, with the highest levels in the caecum (Fig. 6.16). They showed that these acids were present in roughly the same proportions as in the rumen or caecum of a ruminant, or the caecum or colon of a horse.

Friend, Cunningham & Nicholson (1963a, b), Alexander & Davies (1963), Ledinek (1970), Argenzio & Southworth (1974) and Clemens, Stevens & Southworth (1975), examining gut contents of slaughtered pigs, also found VFA in the contents along the entire length of the tract, but all these workers found lactic acid to be the principal organic acid in the stomach, while VFA predominated in the large intestines. The studies of these various authors altogether covered suckled and early weaned piglets from 1 week old, and pigs of all sizes up to 176 kg.

Studies on the gut flora of suckled and early-weaned piglets, growing and mature pigs have shown high levels in the stomach of Lactobacillus and Bifidobacterium species with relatively low levels of other organisms, while the large intestines contain a far more mixed flora (Smith & Jones, 1963; Van der Heyde & Hendrickx, 1964; Ogata & Morishita, 1969; Ledinek, 1970; Decuypere & Van der Heyde, 1972). As was pointed out by Decuypere, Vervaeke & Hendrickx (1974), such a gastric flora will readily convert fermentable carbohydrates to lactic acid, while the mixed flora in the large intestine will produce VFA, mainly acetic acid.

Gastric fermentation

Fermentation products

Friend, Cunningham & Nicholson (1963a, b), studying weaned pigs from 1 week to maturity found that the total organic acids in the stomach amounted to about 150 meq/l in mature pigs and 30–92 meq/l in early weaned piglets. Of this, lactic acid constituted over 90% of the molar concentration for the first three weeks of life on either a whey or a barley diet. Beyond this age, with pigs on a whey diet, lactic acid continued to provide 90% of the gastric organic acid in most pigs of all ages, but on cereal diets, the lactic acid proportion fell to 50%. Propionic acid usually constituted about half of the remaining VFA with progressively lower concentrations of butyric and valeric acids. In the older pigs, formic acid was also present in significant concentrations (3.6–4.5% of organic acids) exceeding those of butyric acid on the cereal diets and those of propionic acid on the whey diet.

Location in the stomach of fermentation products

Slivitskii (1973, 1975), using pigs with a gastric fistula which permitted sampling at 3 different levels in the stomach, and also a re-entrant duodenal fistula, measured the lactic acid, total VFA, free hydrochloric acid, total acidity and pH at three levels in the stomach and in the duodenum of live, conscious pigs at different times after feeding. He found that the lactic acid and VFA were present mainly in the uppermost layer of the stomach. The hydrochloric acid concentration (acidity titratable at pH 4.5) was greatest in the lowest gastric layer which, of course, had the lowest pH, while there was the lowest hydrochloric acid concentration in the uppermost layer, which had the highest pH (Table 6.5).

Table 6.5 *Concentrations of lactic acid, volatile fatty acid and free HCl and the pH at three levels in the stomach and in the duodenum of five 9–10 month old pigs with a gastric fistula and a duodenal re-entrant fistula. (from Slivitskii, 1973, 1975)*

Position sampled		meq/l			pH
		Lactic acid	V.F.A.	Free HCl	
			4 hours after feed		
Stomach	upper	8.6 ± 4.5	10.5	20.0	2.90
	mid	3.5 ± 0.3	7.5	42.0	1.83
	fundal	3.1 ± 0.3	3.6	97.6	1.01
Duodenum		3.1 ± 0.3	3.4	0.0	5.27
			8 hours after feed		
Stomach	upper	31.1 ± 6.6	29.4	15.4	2.78
	mid	16.9 ± 4.5	15.0	65.0	1.64
	fundal	4.8 ± 0.8	5.8	115.1	0.93
Duodenum		5.9 ± 1.5	4.6	0.0	5.84
			10 hours after feed		
Stomach	upper	49.9 ± 8.4	40.0	18.0	2.64
	mid	25.1 ± 8.4	24.8	65.3	1.53
	fundal	5.1 ± 1.7	10.7	100.2	1.22
Duodenum		4.1 ± 0.1	5.7	0.0	5.48
			12 hours after feed		
Stomach	upper	28.3 ± 4.7	44.8	42.8	2.09
	mid	17.0 ± 3.1	25.9	74.0	1.51
	fundal	6.9 ± 0.8	10.4	117.4	0.93
Duodenum		4.7 ± 0.1	7.5	0.0	4.54

Argenzio & Southworth (1974), in samples of contents taken after slaughter, also found higher lactic acid and VFA concentrations in the upper than in the lower half of the stomach.

Variation with time

Étienne (1971) reported a slight progressive rise in lactic acid concentration in the stomach contents after feeding, continuing for 9 hours. Slivitskii (1973, 1975), in fistula studies, showed that the levels of lactic acid and VFA rose throughout the stomach for 10–12 hours after feeding (Table 6.5). Clemens, Stevens & Southworth (1975) examining gastric contents of pigs slaughtered at different times after feeding also found that lactic acid levels were low in pigs killed immediately after feeding and were progressively higher in pigs killed at longer intervals after the feed (Fig. 6.17).

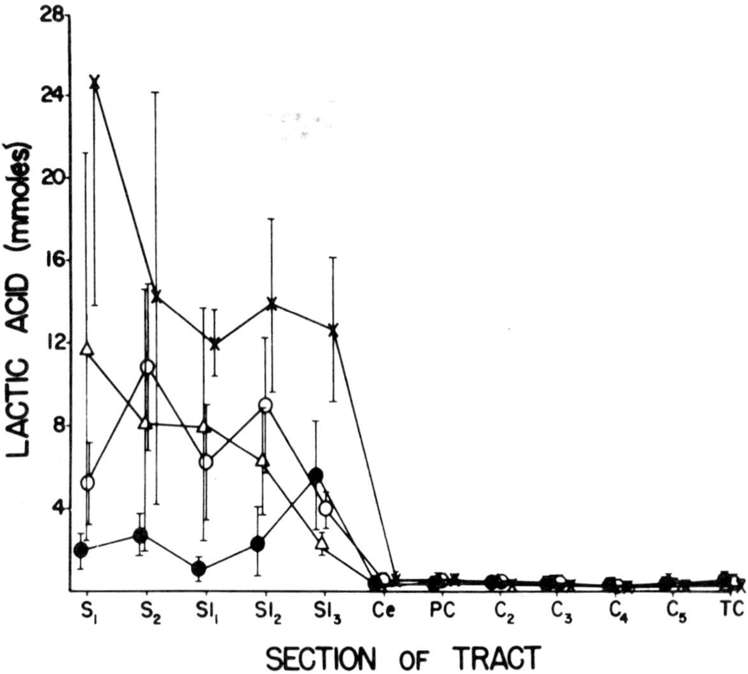

Fig. 6.17 Mean values (± SEM) for the quantity of lactic acid in gastrointestinal contents of the pig. Symbols within the graph correspond to the four time periods: (•), 0 or 12 hours; (△), 2 hours; (○), 4 hours; and (x), 8 hours after feeding. Symbols along the abscissa: S_1, S_2 = stomach; SI_1, SI_2, SI_3 = small intestine; Ce = caecum; PC = proximal colon; C_2, C_3, C_4, C_5 = colon; T. C. = terminal colon (n = 3), (from Clemens, Stevens & Southworth, 1975)

Variation with diet

Friend, Cunningham & Nicholson (1963a, b) did not find a difference in total levels of organic acids on different diets, but found more gastric lactic acid and less VFA on whey diets than on cereal diets. Ledinek (1970) found that when raw potatoes were replaced by cooked potatoes, more fermentation products were found in the stomach and less in the large intestines, the cooked potato providing a medium for gastric fermentation, while the raw potato provided a relatively indigestible starch, a significant amount of which reached the large intestines. Mason & Just (1976), exchanging potato starch for maize starch in pig diets, found that VFA levels in the faeces were higher on the potato starch, more of which reaches the large intestines. On either the maize or the potato starch diet, the inclusion of antibiotics lowered the levels of faecal VFA.

Comparing more widely different diets, Ly & Boucourt (1975) found that with pigs fed a ration containing 64% molasses, the stomach became the main organ of fermentation and contained more VFA than lactic acid, but substituting maize for the molasses, they obtained the same results as most other workers, with less fermentation in the stomach than in the large intestines and the gastric fermentation producing mainly lactic acid.

The Piglet

Cranwell, Noakes & Hill (1968, 1976), Hill, Noakes & Lowe (1970) and Noakes (1972) showed an inverse relationship between lactic acid production and hydrochloric acid secretion in suckled piglets up to the age of 5 weeks. Their interpretation is that, up to this age, the capacity of the gastric fundal mucosa to secrete hydrochloric acid is limited, and sow's milk does not provide a strong stimulus. Under these conditions, the acidity produced by lactic acid from fermentation in the stomach of suckled piglets could be sufficient to suppress hydrochloric acid secretion so that the pH remains high enough for rapid proliferation of lactobacilli but not other organisms, with the production of more lactic acid. They found that this effect was observed where pigs were reared under "conventional" farm conditions where heavy bacterial exposure of the piglets was likely. Where the pigs were reared under "clean" conditions, so that a high population of lactobacilli did not get established in the stomach of the piglets in the first few days of life, hydrochloric acid secretion started and prevented the development of high levels of lactobacilli in the stomach. Figs. 6.18 and 6.19 show the levels of lactic acid and total acidity obtained by Cranwell et al. in a piglet born in a "clean" (Fig. 6.18) and in a "conventional" (Fig. 6.19) environment. They also show that in the suckled piglet, as in the adult, the lactic acid level falls after feeding and then rises before the next feed.

Intestinal fermentation

The small intestine

Friend, Cunningham & Nicholson (1963a, b) found the total organic acid concentration to be lower in the small intestine than in any other part of the alimentary tract. They consistently found formic acid in the small intestine from 3 weeks of age. This was not present in the large intestines, and only present in the stomachs of the larger pigs, and then as a smaller proportion of the organic acids than in the small intestine. They found considerable quantities of lactic acid, which, however, formed a smaller proportion of the total organic acids than was the case in the stomach. The remaining organic acid was mainly acetic with sometimes some propionic.

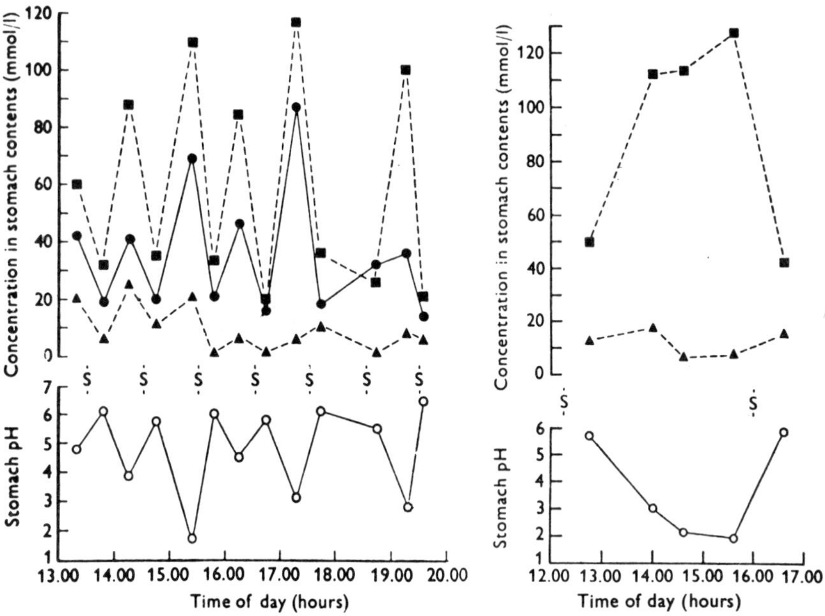

A B

Fig. 6.18 A. Analyses of stomach contents from a 13 day old pig born and reared in a 'clean' environment. (■), chloride; (●), total acidity; (▲), lactic acid; (○), pH; (S), occurrence of suckling.
B. Analyses of stomach contents from a 15 day old pig born and reared in a 'clean' environment, during a 5 hour period in which the pig was not given access to the sow for 4 hours. (■), chloride; (▲), lactic acid; (○), pH; (S), occurrence of suckling. (Cranwell, Noakes & Hill, 1976)

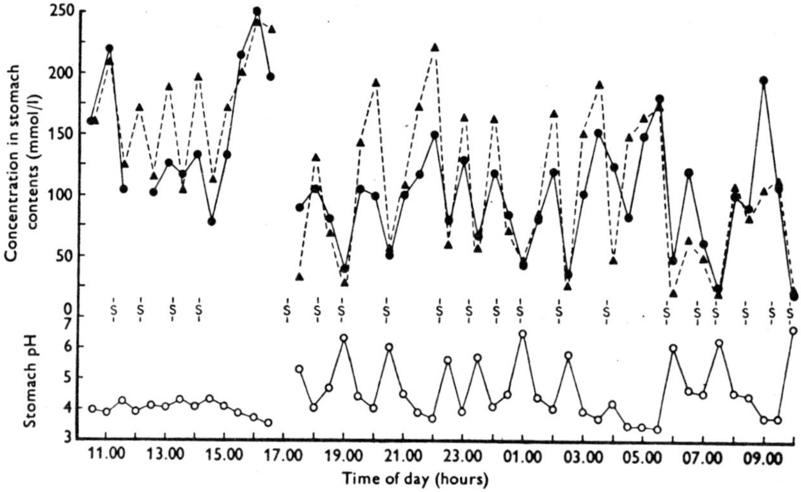

Fig. 6.19 Analyses of stomach contents from a 4–5 day old pig born in a 'conventional' and reared in a 'clean' environment, sampled at 30 minute intervals throughout a 24 hour period. (●), total acidity; (▲), lactic acid; (○), pH; (S), occurrence of suckling. (Cranwell, Noakes & Hill, 1976).

The large intestines

Friend, Cunningham & Nicholson (1963a, b) found organic acid levels of 110–300 meq/l in the caecum and colon, the lower levels being found in the youngest (1 week old) pigs. The main component was acetic acid followed by propionic acid, then butyric acid, with lactic and valeric acids usually being present in only small quantities. They found little difference in levels or proportions between the caecum and the proximal and distal colon. Argenzio & Southworth (1974), with growing pigs, and Clemens, Stevens & Southworth (1975), with mature pigs, found similar levels (150–250 meq/l) of total VFA with little difference between one section of the large intestine and another. Argenzio & Southworth also found that the proportions of the different VFA's were not affected by the position along the large intestine, but were influenced by the diet. They, and Clemens, Stevens & Southworth, also found that, in contrast to the findings with lactic acid in the stomach, the VFA concentrations in the large intestines were not greatly influenced by the time interval between feeding and sampling (Fig. 6.20).

As pointed out in relation to gastric fermentation, Ledinek (1970), Ly & Boucourt (1975) and Mason & Just (1976) have shown that where the diet contains more digestible or fermentable carbohydrate, there will be relatively more fermentation in the stomach and less in the large intestines. Rensmann (1976) found that when 15–40 kg pigs were fed a diet low in fibre, the inclusion of wood pulp in the diet reduced the total faecal VFA concentrations, while

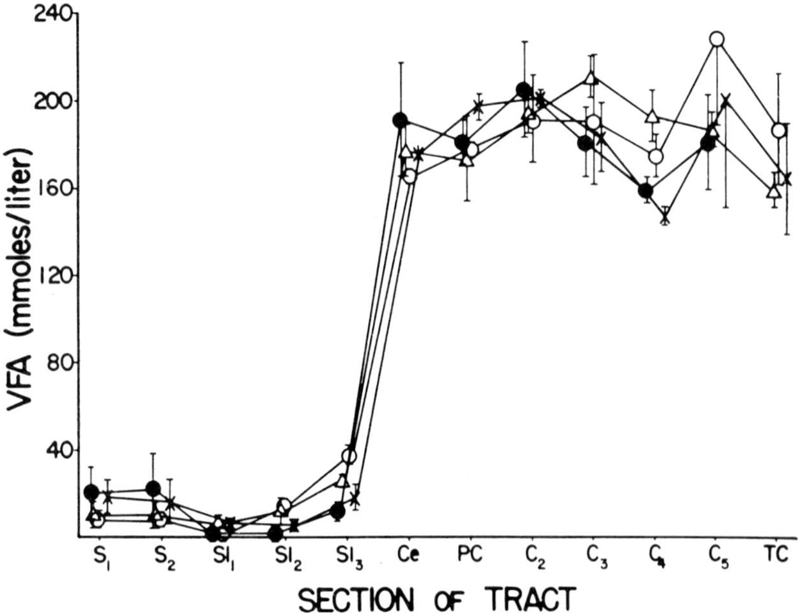

Fig. 6.20 Mean values (±SEM) for the concentration of volatile fatty acids in gastrointestinal contents of the pig. Symbols within the graph correspond to the four time periods: (•), 0 or 12 hours; (△), 2 hours; (○), 4 hours; and (x), 8 hours after feeding. Symbols along the abscissa are the same as used in Fig. 6.17 (n = 3). (from Clemens, Stevens & Southworth, 1975)

Table 6.6 Transport of VFA across porcine gastrointestinal mucosa (Argenzio & Southworth, 1974)

Gastrointestinal mucosa	No. of Expts.	Volatile fatty acid, µmol/cm²		
		Loss lumen side	Gain blood side	Tissue content
Gastric stratified squamous	5	15.2 ± 6.0	1.0 ± 0.2	2.8 ± 0.6
Cardiac	5	20.8 ± 4.6	2.8 ± 0.7	2.0 ± 0.4
Proper gastric	5	12.6 ± 2.5	0.8 ± 0.1	2.2 ± 0.4
Pylorus	4	14.6 ± 2.7	1.3 ± 0.5	5.6 ± 0.7
Caecum	5	25.8 ± 7.3	10.7 ± 1.6	2.8 ± 0.2
Centripetal colon	4	20.1 ± 1.9	9.3 ± 1.6	2.8 ± 0.4
Centrifugal colon	5	24.4 ± 2.4	7.7 ± 0.6	3.2 ± 0.2

Values are means ± SE of results obtained during a 2.5 h experimental period inclusion of potato starch increased it. The concentrations or proportions of the faecal VFA's were not appreciably influenced by age, but the wood pulp in the diet increased the acetic and propionic acids and reduced the butyric acid. Potato starch in the diet increased the faecal propionic and butyric acids.

Argenzio & Southworth showed that, in vitro, large intestine mucosa (caecum, centripetal colon and centrifugal colon) had a high capacity to transport VFA from the mucosal to the serosal side. This contrasted with the low transport capacity shown by the various types of gastric mucosa (Table 6.6).

QUANTITATIVE ASPECTS OF CARBOHYDRATE DIGESTION

Sugars

The main energy source of diets of adult pigs in most of the world is starch rather than sugars, starch sources usually being cheaper. Sugar, if present in the diet of mature pigs, has usually been added in the form of spent beet pulp, whey, or molasses, the latter used for binding a pelleted ration. It is, therefore, not likely to be present at a high level. As common sugars other than lactose are efficiently digested and absorbed by the adult pig, the relatively limited quantities in a usual pig diet are likely to be completely digested and absorbed. For this reason, the digestion of sugars by the adult pig has attracted little interest or study outside the major sugar producing countries.

In the very young pig, where the capacity for starch digestion is rather limited and consequently high levels of sugar may be included in the ration, the situation is quite different, and a great deal of work has been done related to the suitability or otherwise of different sugars in early weaning diets.

Digestion of sugars in piglets

Comparisons of survival and growth of piglets on diets containing high levels of different sugars have been made by a number of workers, whose results, where they were studying the same sugars, were in agreement with one another. Their conclusions as to which were satisfactory dietary sugars for piglets are summed up in Table 6.7. Where a sugar was unsatisfactory, there was scouring, poor growth and some mortality. Where it was very unsatisfactory there was severe scouring and high mortality. The results in the disaccharides are in good agreement with the enzyme measurements already described, where maltase, and particularly, sucrase, are very low at birth and rise with age, while lactase is very high at birth and declines with age.

Further information has been obtained by feeding sugars or infusing them into the alimentary tract and then measuring either the quantity remaining in the lumen, or the rise in blood level. Kidder, Manners, McCrea & Osborne (1968) fed piglets on diets containing different sugars with a non-absorbable marker, slaughtered the piglets at 1, 2 or 3 weeks of age, 2½ hours after a meal and measured the sugar and marker in the gut contents. Glucose had always disappeared from the lumen by the second half of the small intestine. Fructose and xylose were more or less equally, and far less completely absorbed. Sucrose was still less well absorbed, especially by the younger pigs, and was less completely absorbed when it formed 15% of the diet than when it formed 5% (Fig. 6.21). They found a considerable quantity of sucrose present in the large intestine on the high sucrose diets, producing water retention, confirming the supposition of

Table 6.7 *Relative suitability of various carbohydrates for piglets*

Sugar	Response	Reference
Glucose	Very satisfactory at any age	1, 2, 3, 4
Fructose	Unsatisfactory for very young piglets	1, 4
Xylose	Unsatisfactory for 2 week old piglets	5
Lactose	Very satisfactory for very young piglets	1, 3
	Less satisfactory for piglets of several weeks of age	1, 2
Maltose	Satisfactory at any age but not as good glucose for very young piglets	6
Sucrose	Extremely unsatisfactory for very young piglets	1, 3
	Less unsatisfactory with increasing age	3, 7
	Some individuals adapt early	3
	Satisfactory at 9 weeks	2

References
1) Johnson, 1949
2) Becker & Terrill, 1954
3) Becker, Ullrey & Terrill, 1954
4) Becker, Ullrey, Terrill & Notzold, 1954
5) Wise, Barrick, Wise & Osborne, 1954
6) Cunningham & Brisson, 1957
7) Sewell & Maxwell, 1966

Becker & Terrill (1954) that this was the cause of the ill-effect of sucrose in piglet diets.

Cunningham & Brisson (1957) injected maltose into tied off intestinal loops of piglets aged 0–7 days, and removed the fluid 1 hour later. They found that maltose was digested at a rate of 0.62 g/kg body wt/hour from birth to 4 days and 0.99 g from 6–7 days of age. In a feeding trial over the first week of life, they found maltose to have an apparent digestibility of 97.4% as against a value for glucose of over 99%.

Rises in blood sugar level after feeding a test sugar or infusing it into the lumen of the gastrointestinal tract, have been used for comparison of rates of intestinal absorption of different sugars, or their digestion products.

Dollar, Mitchell & Porter (1957) compared the blood sugar rises after oral doses of glucose, lactose, maltose and sucrose in piglets of 2–4 days, 9–11 days and 15–22 days of age. They found large rises after glucose, which were similar for each age group, and small rises with the other sugars, those after lactose decreasing with age, and those after maltose and sucrose increasing. In the cases of both the sucrose and the maltose, there was no significant rise in the youngest pigs (Fig. 6.22).

Cunningham (1959), in studies on starch digestion referred to later, compared blood sugar rises after glucose and maltose infusion. He found that in 3 day old pigs there was only a small rise in blood sugar after maltose infusion, but

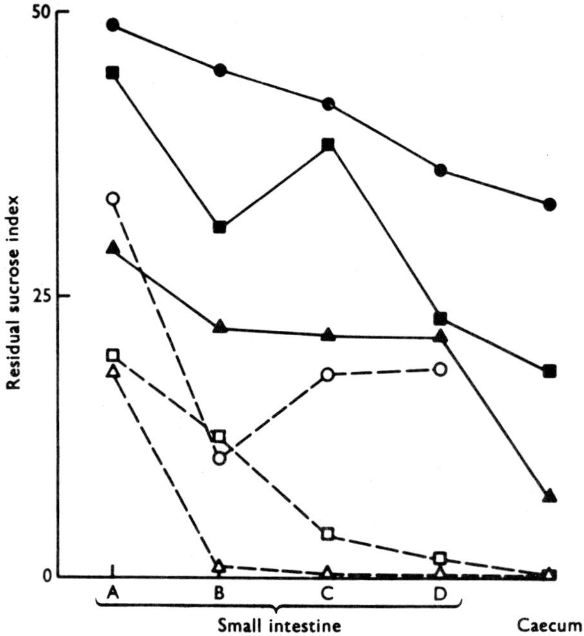

Each point is the mean of results for two piglets. ○, 1 week old, low (5 %) sucrose diet; □, 2 weeks old, low (5 %) sucrose diet; △, 3 weeks old, low (5 %) sucrose diet; ●, 1 week old, high (15 %) sucrose diet; ■, 2 weeks old, high (15 %) sucrose diet; ▲, 3 weeks old, high (15 %) sucrose diet.

Fig. 6.21 The ratio of sucrose to marker in the gut contents as a percentage of the dietary ratio in the four quarters (by length) of the small intestine and in the caecum of piglets on diets containing different levels of sucrose. (from Kidder, Manners, McCrea & Osborne, 1968)

a large rise after glucose, while with pigs of 7 days of age or more, glucose and maltose gave similar rises (Figs. 6.27 & 6.28).

Kidder, Manners & McCrea (1963) compared the blood fructose levels after feeding sucrose or invert sugar as 20% of the diet to pigs aged 5, 10 or 17 days. The rises were greater on the invert sugar than on the sucrose at 5 days of age and, to a lesser extent, at 10 days but not in the 17 day old pigs.

It is clear that the results of studies on sugar digestion in piglets, using as criteria survival and growth, disappearance of the sugar from the gut lumen, or appearance of the sugar in the peripheral blood, all agree in general with one another and with the findings on the mucosal enzyme levels. Obviously the different techniques will be more sensitive indicators under different circumstances.

Means for three litters; ●——●, 2–4 days; △ - - - △, 9–11 days; ○—·—○, 15–22 days

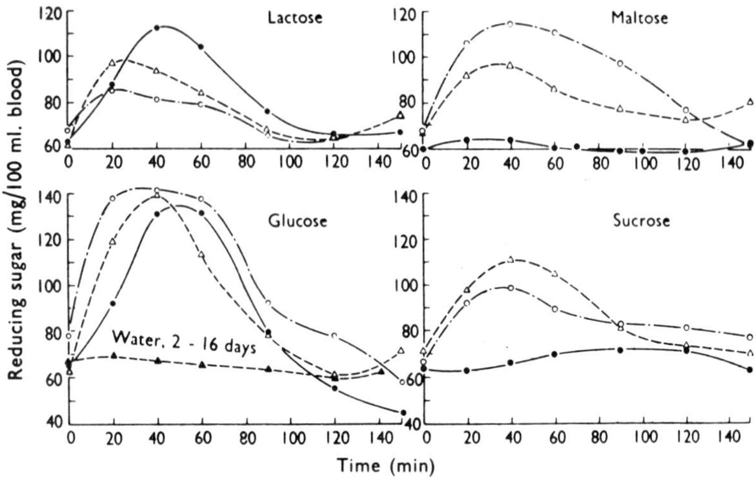

Fig. 6.22 Blood sugar levels in piglets of different ages at various times after an oral dose of 0.82 g/lb of different sugars. (from Dollar, Mitchell & Porter, 1957)

For example, a difference between 85% and 95% absorption in the small intestine will scarcely be reflected in the blood level, but will make a 3-fold difference to the residual quantity of sugar in the distal ileum. Table 6.7 on the relative suitability of various sugars in the diet of piglets is therefore equally applicable to their digestibility.

Digestion of sugars in adult pigs

PERFORMANCE ON DIFFERENT SUGARS

Lactose

Becker, Terrill, Jensen & Hanson (1957) found that, while with 14-day-old piglets, 60% dried whey (giving 42% lactose in the diet) could be included with no detrimental effects, in 8 week old (40 lb) pigs 30% dried whey was detrimental, producing scouring and depressed growth. With 4 month old (100 lb) pigs, 60% whey depressed growth and caused scouring, while 40% caused some scouring. Dunkin (1959) reported scouring in pigs on liquid whey diets. Shearer & Dunkin (1968a) fed lactose as an alternative to starch at 0, 15, 30 & 45% of the diet to 50 lb pigs, and found that the higher the lactose level, the worse was the performance, and at the 45% level, some pigs scoured throughout the trial. On the other hand, Ekstrom, Benevenga & Grummer (1975c) found that feeding up to 40% whey (about 30% lactose in the diet) in a maize-soya ration on pigs up to 21 weeks of age actually improved performance and feed utilization, except in two pigs which were found at slaughter to have low intestinal lactases. These authors concluded that growing pigs could tolerate up to 30% lactose in the diet well. In other experiments, Ekstrom, Grummer & Benevenga (1976) found that 40% dried whey in the diet caused scouring in a number of pigs but only caused poor performance in the Hampshires, not in the Chester Whites. The lactase levels of the intestinal mucosa of their Hampshires was about half that of their Chester Whites. In neither breed was the mucosal lactase level influenced by diet, but in both breeds, the lactase of the caecal contents, presumed to be due to the bacterial flora, was higher on the whey diet. They showed that lactose from the whey reached the caecum, and Kim, Benevenga & Grummer (1976) showed that adaptive changes of the caecal flora occurred when whey was fed from 12 weeks, and that this improved subsequent performance on a high lactose diet.

Shearer & Dunkin (1968b) showed that pigs fed up to 120 lb on high lactose diet had a larger caecum than those on a control diet. Ekstrom, Benevenga & Grummer (1975b) showed that with 21 week old pigs on diets containing high lactose levels, significant quantities of lactose could be found in the caecum, having passed through the small intestine unabsorbed. This would presumably retain water and increase the volume of the material passing through the caecum. These reports are consistent with the fall in level of intestinal lactase with age already described.

Sucrose

Molasses has been used as a major dietary constituent in sugar-producing countries (e.g. Brooks & Iwanaga, 1967), and has been found to produce diarrhoea and poor performance when used at levels of more than 30% (Blanco, Raun & Vargas, 1964). Macleod, Preston, Lassota, Willis & Velasquez (1968) have shown that this is not due to the sucrose content. Replacement of part of the molasses by crude cane sugar abolished the soft faeces and produced good growth and energy conversion in pigs through the age range 20–90 kg. In these experiments, the best results were obtained with the greatest replacement of molasses by sugar, giving diets in which sucrose constituted 67–70% of the dry matter. In a similar experiment, Velasquez, Ly & Preston (1969) showed that, as the molasses was replaced by crude sugar so that the sucrose content rose, the faecal dry matter increased, with the result that a diet containing 62.7% sucrose gave a faecal dry matter of 40.7%. Thus the capacity of the pig of 20 kg upwards to digest sucrose is very high.

ABSORPTION OF SUGARS

Sugar absorption in the intestine can only be studied on a starch-free diet, otherwise sugar is being continuously produced from the starch at an unknown rate. A few studies have been made on sugar absorption in adult pigs using isolated intestine loops or with re-entrant ileal fistulas.

Using the intestinal loop technique, Kvasnitskii (1951) and Furuya & Takahashi (1975) both measured glucose disappearance from the lumen, in each case on three pigs with isolated jejunal loops. Kvasnitskii (1951) used 6–9 month old pigs with intestinal loops 30–35 cm long and 2.5–5 metres from the stomach, and 10 minutes after infusion of glucose solution drained and washed out the contents of the loop. His results (Table 6.8) show that the intestinal loop responds to increased nutrient either in the loop itself or in the main part of the intestine by increased absorptive activity, giving a range of absorption from about 5 to 114 mg per metre of intestine per minute (0.3–6.8g/h). Furuya & Takahashi (1975), using 10–15 week old pigs, 25–30 kg liveweight, perfused a 1 m section

Table 6.8 *Quantity of glucose (in mg) absorbed in 10 minutes from an isolated intestinal loop in a pig under various conditions. (from Kvasnitskii, 1951)*

	Concentration of glucose solution					
	0.5%		1.0%		3.0%	
Feeding regime of pig	Day	Night	Day	Night	Day	Night
Fasted 72 hours	27.4	18.5	57.2	30.0	88.0	47.0
Normal feeding	28.1	24.7	64.7	68.1	144.7	125.5
Heavy feeding	24.5	32.5	58.8	63.0	336.9	380.6

of jejunum from one fistula to another with solutions containing glucose and P.E.G. with or without L-histidine. The rate of absorption of glucose rose from 1 to 3 g per hour with increased glucose concentration, but the rise was not proportional to concentration and was unaffected by the presence or absence of histidine. Increasing the flow rate from 400 to 800 ml/hour decreased the percentage of glucose absorbed, and produced little or no increase in amount absorbed. They also found that the absorption was less at night than during the day (Table 6.9).

Table 6.9 *Effect of period of measurement on the absorption of L-histidine and glucose from a test solution containing 10 g of each per l, perfused at a rate of 600 ml/l through a jejunum segment of pigs fitted with three simple cannulae (from Furuya & Takahashi, 1975)*

(Mean values with their standard errors for three pigs)

	Absorption (%)				Rate of absorption (g/h)			
Period of measurement (hours)	L-histidine		Glucose		L-histidine		Glucose	
	Mean	SE	Mean	SE	Mean	SE	Mean	SE
10.0–10.30	36.7	2.2[a]	60.2	5.3[a]	2.20	0.10[a]	3.61	0.19[a]
16.00–16.30	34.3	2.0[a]	52.7	5.5[a]	2.06	0.10[a]	3.16	0.29[a]
22.00–22.30	29.2	1.8[b]	45.2	5.1[b]	1.75	0.09[b]	2.71	0.26[b]

a, b Values in the same column with different superscripts are significantly different ($P < 0.01$)

Feeding pure sugar solutions to two 30 kg pigs with re-entrant ileal fistulas, Cunningham, Friend and Nicholson (1963) found that the digestibility up to the terminal ileum was greatest with glucose, followed by galactose, then sucrose and very much less with lactose, as would be expected from performance results.

Ly (1971) fed a diet based on molasses and yeast, and containing 75% sugars on the dry matter, to 40 kg pigs with re-entrant fistulas in the terminal ileum, and obtained digestibilities by the terminal ileum, of 97.1–99.8% for the total sugars. Thus, although it is not as well absorbed as glucose, sucrose is nevertheless very largely absorbed in the small intestine even when fed at this very high level. Ly (1975) found considerable variation through the day in the caecal contents, so that on a high molasses diet the sugar level in the organic matter of the caecal contents varied from 9–30%. About half this sugar in the caecum was fructose and about a third was sucrose.

Comparisons of rate of absorption of different carbohydrates in adult pigs have also been made using blood sugar levels as an index. Ly & Velasquez (1970) compared a molasses diet containing 63% sucrose, a molasses diet containing mainly hexoses, and a maize starch diet. These were fed to 30 kg pigs at 5% of body weight and the peripheral blood sugar levels monitored. The hexose diet gave the highest blood sugar, then the sucrose, while the maize starch gave least. After 3 hours the peripheral blood levels were no longer different on the different diets. Rérat, Février, Aumaitre, Vaissade & Vaugelade (1973) compared portal and jugular blood sugar levels after feeds of glucose, sucrose, lactose and maize starch, 400 g being fed to 30 kg pigs. Glucose was most rapidly absorbed, giving the most rapid rise in portal blood sugar level and the most rapid disappearance of the difference between the portal and jugular levels (Figs. 6.23, 6.24, 6.25 & 6.26). Lactose was clearly the least rapidly absorbed, with the least rise in the portal blood level, while the difference between the portal and peripheral blood sugar level, small as it was, had hardly diminished after 8 hours.

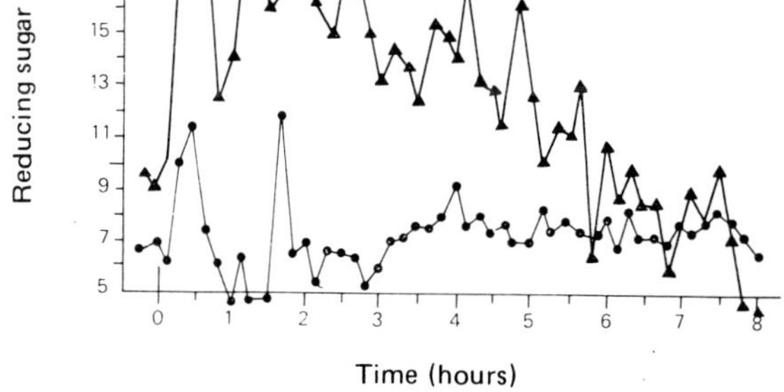

Fig. 6.23 Portal and jugular blood sugar levels in the pig before and after ingestion of glucose. (from Rérat, Fevrier, Aumaitre, Vaissade & Vaugelade, 1973)

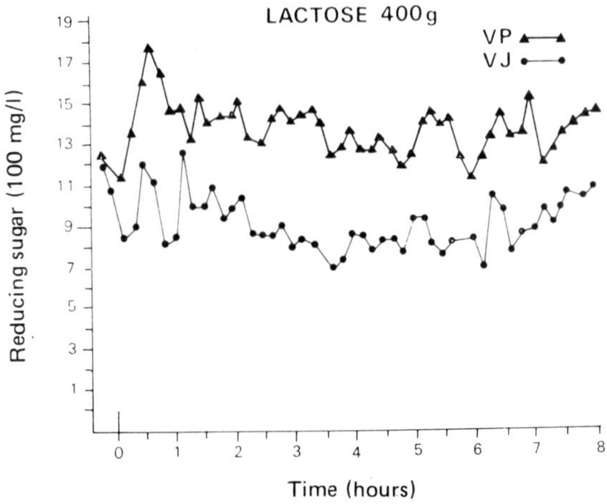

Fig. 6.24 Portal and jugular blood sugar levels in the pig before and after ingestion of lactose. (from Rérat et al., 1973)

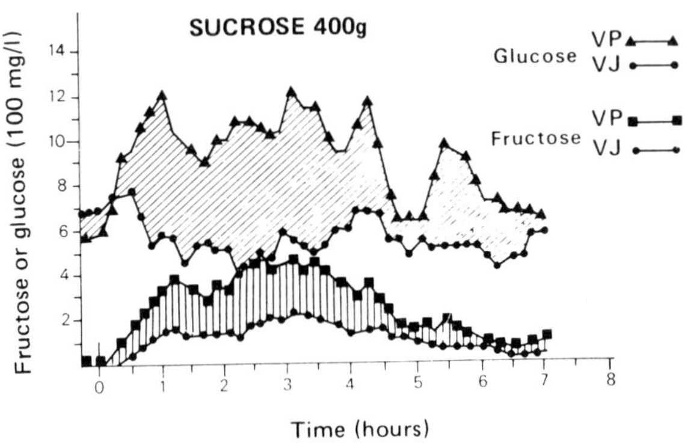

Fig. 6.25 Portal and jugular blood glucose and fructose levels in the pig before and after ingestion of sucrose. (from Rérat et al., 1973)

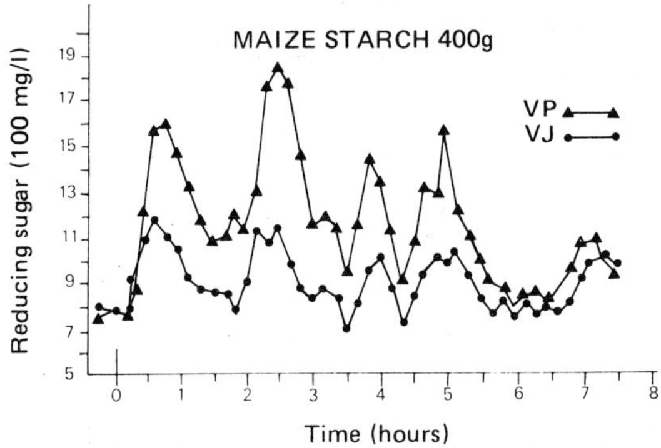

Fig. 6.26 Portal and jugular blood sugar levels in the pig before and after ingestion of maize starch. (from Rérat et al., 1973)

Starch

Digestion of starch has been studied to a limited extent in piglets and more extensively in older pigs.

Starch digestion in piglets

Starch is not well digested by very young piglets and a number of studies have been made with a view to determining the earliest age and the maximum level at which it can safely be included in early weaning diets.

GENERAL PERFORMANCE

Becker, Ullrey & Terrill (1954) found that piglets of 5–35 days did equally well for weight gain and survival on diets containing 56% of either glucose, lactose, dextrin or corn starch. Cunningham (1959), comparing glucose, maltose, soluble and raw corn starch on piglets of 3–10 days and 10–23 days of age, found that in the earlier period the daily weight gains were 71, 51, 48 and 32 g on the glucose, maltose, soluble and raw starch respectively. The digestibilities of the four carbohydrates were similar and all 97% or above, the difference in performance being related to voluntary food intake.

Sewell & Maxwell (1966), with pigs from 3 weeks old, found dextrin to be inferior to glucose, lactose, sucrose and corn starch, while corn starch was inferior to the sugars if given as the only carbohydrate.

ABSORPTION

Cunningham (1959) injected raw soluble starch with or without amylase into intestinal loops of piglets of 20 hours and 5, 10 and 15 days of age and measured the amount of starch disappearing in 2 hours. He found that both types of starch, but especially the raw starch, were poorly digested in the youngest pigs, but that the amount digested increased with age (Table 6.10).

Table 6.10 Effect of amylase and intestinal digestion of raw and soluble starch by baby pigs (Cunningham, 1959)

	Raw starch				Soluble starch			
	0 days	5 days	10 days	15 days	0 days	5 days	10 days	15 days
	Total starch injected per pig							
	1.84	4.60	6.90	9.20	1.97	4.93	7.39	9.86
	Rate of digestion (gm/kg. body wt./hr.)							
	Controls							
Rep.1	0.10	0.47	0.54	0.50	0.37	0.44	0.81	0.65
Rep.2	0.70	0.35	0.38	0.71	0.47	0.71	0.56	0.61
Av.	0.08	0.41	0.46	0.60	0.42	0.57	0.68	0.63
	Amylase							
Rep.1	–.02	0.27	....	0.42	0.42	0.88	1.13	....
Rep. 2	0.15	0.58	0.29	0.45	0.50	0.64	0.51	1.02
Av.	0.06	0.42	0.29	0.43	0.46	0.76	0.82	1.02

Braude, Dollar, Mitchell & Porter (1958) looked at the blood sugar rises after feeding various starch products to piglets. Cooked dextrin gave a marked blood sugar rise in 5–7 day old piglets, raw dextrin at 8–12 days, cooked potato starch at 13–17 days and raw wheat starch at 23 days.

Cunningham (1959) compared the rises in blood sugar level after administration by stomach tube of glucose, maltose, soluble starch and raw starch to pigs 3–42 days of age. He found that, at 3 and 7 days, the starches gave no detectable rise in blood sugar, but by 14 days, soluble starch gave a marked rise, though less than that from glucose or maltose, and that by 42 days, raw starch also produced a rise in blood sugar (Figs. 6.27 & 6.28).

Digestion of starch by adult pigs

Starch is not only the main carbohydrate in most diets for adult pigs, it is also the main energy source. It is well digested, but, because it is a major ingredient, the exact degree to which it is utilized is of economic importance. The "nitrogen free extractives" (N.F.E.), a parameter estimated in the proximate analysis of feedingstuffs, measures primarily the starch and sugar content, so, in a feedstuff containing little or no sugar, it is an approximate measure of starch content. (Chapter 9). The digestibility of N.F.E. has been measured in pigs on a variety of diets. In addition, some more specific measurements have been made on starch.

Although digestion and absorption of dietary starch is not as complete by the end of the small intestine as is that of soluble sugar, starch is nevertheless very largely digested and absorbed by this point. Holmes, Bayley & Horney (1973) showed that, on maize diets, the net digestibility of starch was 94–98% by the end of the small intestine and 100% by the end of the large intestine.

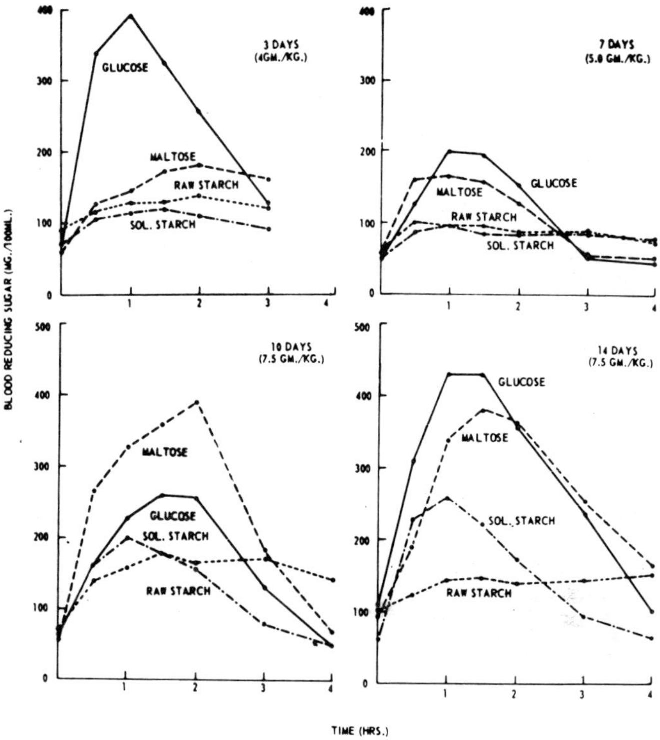

Fig. 6.27 Blood sugar levels of pigs of 3 to 14 days of age given various carbohydrates by stomach tube. The dosage level of carbohydrate is given under the age of each group. (from Cunningham, 1959)

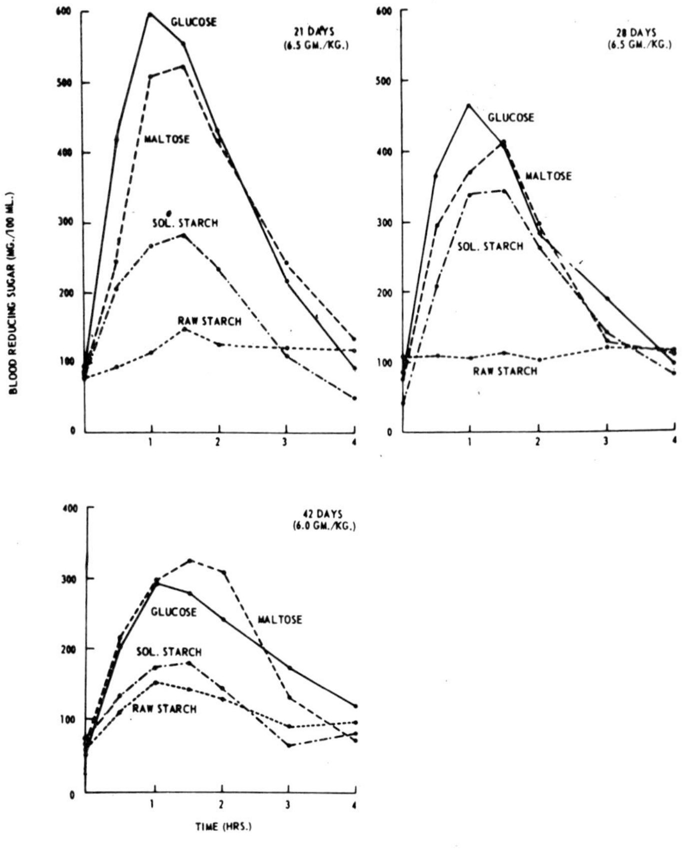

Fig. 6.28 Blood sugar levels of pigs of 21 to 42 days of age given various carbohydrates by stomach tube. The dosage level of carbohydrate is given under the age of each group. (from Cunningham, 1959)

Keys & DeBarthe (1974a) studied the extent of digestion of starch, amylose, amylopectin and sugar in different sections of the gut using fistularized 80–90 kg pigs on different cereal diets. They found that the greater part of the starch, amylose and amylopectin had disappeared by the duodenal fistula, having been substantially converted to sugar. This had largely disappeared by the terminal ileum. Table 6.11 gives the apparent digestibilities of these carbohydrates as found by Keys & DeBarthe.

Table 6.11 Carbohydrate digestion coefficients[a] in swine fed wheat, milo, corn or barley diets (Keys & DeBarthe, 1974a)

Item	Treatment				Standard error	Coefficient of variability	Treatment significance
	Wheat	Milo	Corn	Barley			
Starch							
Duodenum[b]	75.72	63.27	71.83	45.11	4.86	15.20	P < .05
Ileum[b]	94.97	86.90	93.36	79.14	5.09	11.49	N.S.
Faeces[c]	98.46	94.66	98.65	93.57	0.81	1.67	P < .01
Amylose							
Duodenum[b]	95.95	90.05	94.30	69.37	4.07	9.31	P < .05
Ileum[b]	97.53	91.43	96.61	85.28	4.42	9.53	N.S.
Faeces[c]	98.62	94.49	98.44	93.59	0.86	1.79	P < .05
Amylopectin							
Duodenum[b]	70.45	52.20	66.52	40.24	7.66	26.72	N.S.
Ileum[b]	94.30	86.06	92.68	77.74	5.33	12.16	N.S.
Faeces[c]	98.41	94.70	98.67	93.57	0.80	1.67	P < .01
Sugar							
Duodenum[b]	−86.04	−43.42	−405.79	41.14	103.30	167.23	N.S.
Ileum[b]	98.58	99.34	91.53	97.87	1.24	2.57	P < .05
Faeces[c]	99.77	99.79	99.35	99.89	0.11	0.21	P < .05

a Mean of four values

b Determined by Cr_2O_3 indicator

c Determined on total collection basis

Ledinek (1970) showed that in 25–27 kg pigs with caecal fistulas, on a raw potato diet, starch formed one third of the dry matter in the caecal contents, but on cooked potatoes, it formed only one eighth.

Although starch is almost completely digested and absorbed by the end of the small intestine, its absorption is considerably slower than that of glucose, so that, when fed to a pig, it gives a smaller and more prolonged rise in blood sugar. This was very well illustrated by the experiments of Aumaitre, Rérat Vaissade & Vaugelade (1973), who fed glucose and also various quantities of maize starch to pigs with portal vein cannulas, and measured the glucose levels in the portal and

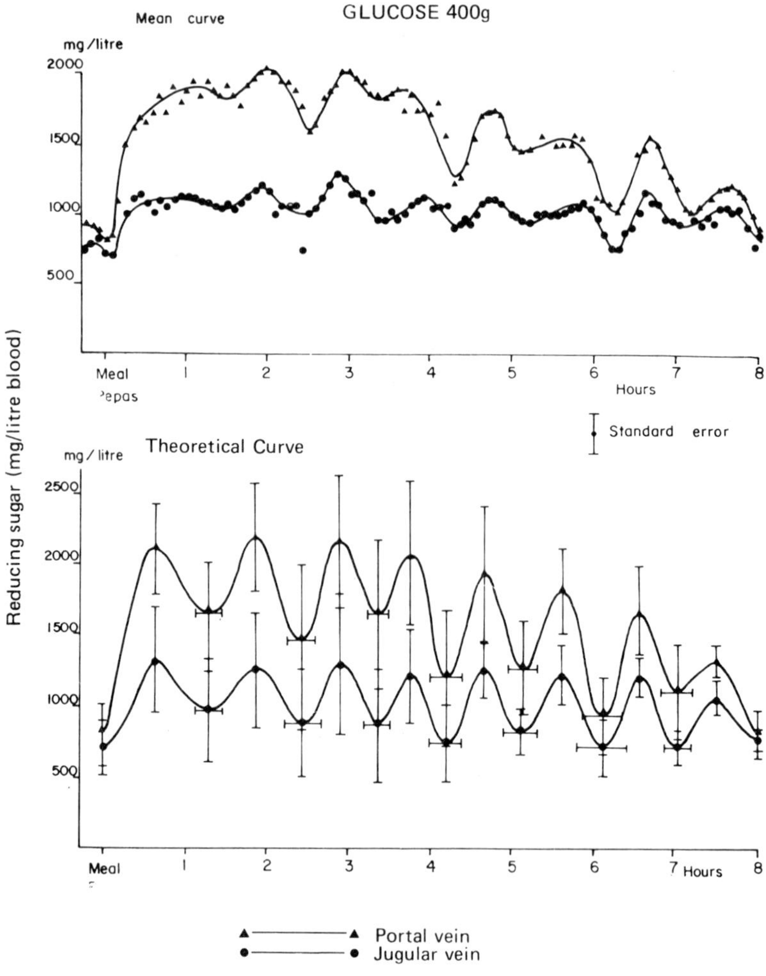

Fig. 6.29. Reducing sugar levels in the portal and jugular blood of a pig after a feed of 400g glucose. Above: actual levels, below: derived theoretical curve. (from Aumaitre, Rérat, Vaissade & Vaugelade, 1973)

jugular blood. The glucose level in both portal and jugular blood rose after the feed, the portal more rapidly, and then both levels showed rhythmical rises and falls (Figs. 6.29 & 6.30). The rises after 400 g of glucose were greater than after 400 g starch, but 8 hours after the glucose feed, the portal and jugular glucose levels were the same, while 8 hours after the same quantity of starch, the portal glucose level was still significantly above the jugular. With larger starch feeds, the initial blood glucose peak was greater, and the difference after 8 hours, much larger (Fig. 6.31).

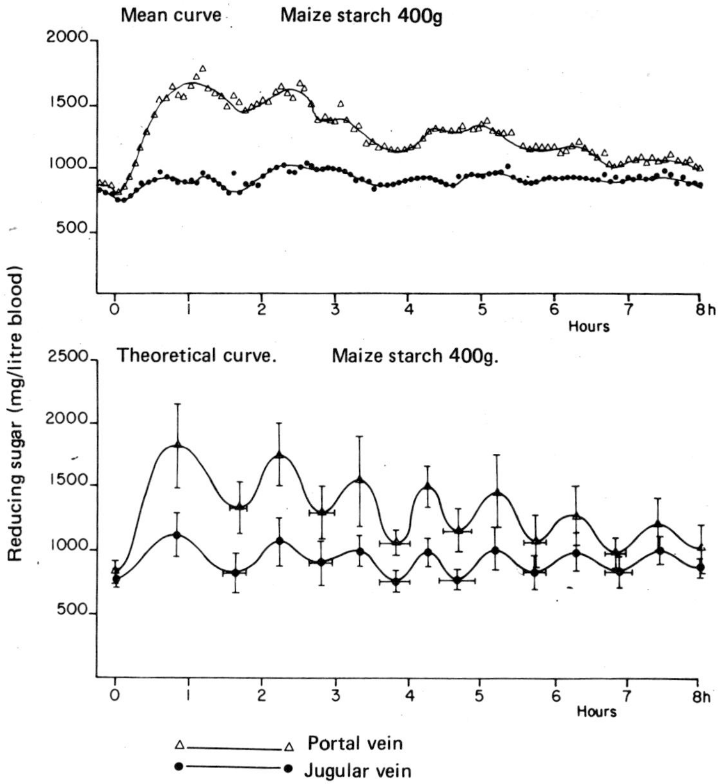

Fig. 6.30 Reducing sugar levels in the portal and jugular blood of a pig after a feed of 400g starch. Above: actual levels, below: derived theoretical curve. (from Aumaitre et al., 1973)

Rérat, Aumaitre, Vaugelade & Vaissade (1974) showed that 16 hours after a feed containing 400 g of maize starch, starch absorption was still continuing, the glucose level of the portal blood being 130 mg/l higher than the peripheral. After 22 hours this difference had disappeared. They estimated that the amount of glucose absorbed by the gut in the period from 16 to 22 hours after the feed was more than 100 g.

In similar experiments, Anderson (1974) found that the portal plasma glucose level rose more rapidly, and to a higher level after a feed of soluble starch than after potato starch, but that the rise after potato starch was more prolonged, indicating a slower digestive breakdown.

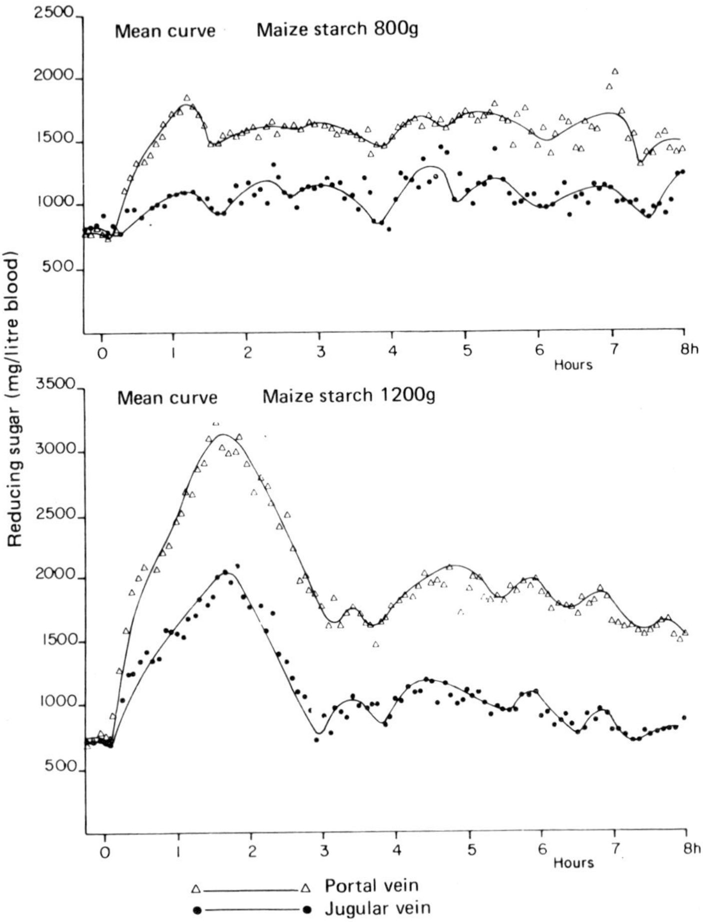

Fig. 6.31 Reducing sugar levels in the portal and jugular blood of a pig after feeds of 800g (above) and 1200g (below) of starch. (from Aumaitre et al., 1973)

Cellulose

Although digestion of true cellulose in the pig only occurs in the large intestine, where it is partly broken down by micro-organisms with the production of volatile fatty acids (Cranwell, 1968), some hemicellulose is broken down by micro-organisms in the stomach. The majority of measurements of cellulosic materials have been measurements of digestibility of "crude fibre", a term which is not specific, but is primarily a measurement of cellulose (see Chapter 9). Some more specific measurements have been made on cellulose.

Horszczaruk and Śljivovaćki studied the rate of breakdown of cellulose in the large intestines of the pig by putting commercial ground cellulose or ground lucerne meal in silk bags in the caecum or colon of fistularized pigs and measuring the amount of "crude fibre" left in the bags after 2 or 4 days. Their findings (Horszczaruk, 1962; Horszczaruk & Śljivovaćki, 1966, 1971) were:

1) The age of the pig was important. More fibre was broken down in the intestines of 160–170 kg pigs than in 120 kg animals, while in 45–50 kg animals practically no ground cellulose disappeared, and relatively little fibre from the lucerne.

2) More crude fibre disappeared when the bag was in the caecum than when it was in the colon.

3) The ground cellulose disappeared to a greater extent than the lucerne fibre in the older pigs.

The rate of disappearance ranged from about 70% in 4 days for ground cellulose to 20% in 4 days for the fibre of unprocessed lucerne meal, processed lucerne meal being intermediate.

4) Ground cellulose disappeared more rapidly if the pig was on a high fibre diet, suggesting adaptation by the caecal flora.

Farrell (1973) measured the apparent digestibility of cellulose and total cell-wall constituents in 4 fistularized 50–60 kg pigs. On diets containing fodder cellulose (alkali-treated straw) as a supplement to the basal commercial ration he obtained apparent digestibilities of just below 50% for cellulose and cell-wall constituents (Table 6.12). Using ground lucerne as a supplement he obtained similar apparent digestibility values for cellulose, but at very high lucerne levels, lower values for cell wall constituents (Table 6.13). Keys and DeBarthe (1974b), making similar measurements with 4 fistularized 82–90 kg pigs on diets containing added hay of four types with different cellulose: hemicellulose ratios, obtained overall digestibilities for the cellulose of 20–48% and, for the hemicellulose, of 25–54% (Table 6.14) but no correlation of digestibility of either with cellulose: hemicellulose ratio. In only one of the four diets did digestion of cellulose begin before the large intestine, but in three of the four, the digestion of hemicellulose had started in the stomach.

The rather wide range of digestibilities found for cellulose and hemicellulose is probably due, as suggested by Farrell (1973), to differences in the nature of the cellulose from different sources or after different treatments, making the material more, or less, susceptible to bacterial action.

The effect of dietary antibiotics on digestibility of cellulose has not been specifically measured, but Mason & Just (1976) obtained striking reductions in digestibility of crude fibre when antibiotics were included in the diet. They also obtained marked reductions in crude fibre digestibility by replacing maize starch in the diet by the less digestible potato starch which provides the large intestine flora with an alternative energy source in place of cellulose.

Table 6.12 *Apparent digestibility of nutrients in the basal diet and in the fodder cellulose (Farrell, 1973)*

	Coefficient of apparent digestibility (%)				
Pig no.	Dry matter	Crude protein	Cellulose	Cell-wall constituents	Energy
	Basal diet				
2	79.5	86.8	38.6	53.8	80.8
4	79.3	86.9	45.6	51.5	80.6
	Fodder-cellulose doet				
1	66.3	76.7	49.7	48.8	68.6
3	66.7	76.8	47.8	47.1	69.1

Table 6.13 *The mean values with SE for digestibility coefficients, for cell walls and for cellulose in diets containing 0 (A), 25 (B), 50 (C), 75 (D) and 100 (E) % ground lucerne. (Farrell, 1973)*

		Diet				
		A	B	C	D	E
Cell walls	Mean	45.0^{a*}	50.5^{b}	46.7^{b}	36.2^{c}	33.5^{c}
	SE	0.8	1.7	0.8	1.1	2.0
Cellulose	Mean	47.2^{a}	54.3^{a}	48.4^{a}	47.2^{a}	51.2^{a}
	SE	1.8	2.8	3.4	0.9	1.2

*Values on a line with different superscripts are significantly different ($P < 0.05$)

Table 6.14 ***Digestibility of cellulose and hemicellulose. Digestibility coefficients[a] of dry matter and cell wall constituents sampled from the duodenum, ileum and faeces of swine fed four hay diets. (Keys & DeBarthe, 1974b)***

	Treatment						
Item	Alfalfa	Grain sorghum	Texas Kleingrass	Coastal Bermuda grass	Standard error	Coefficient of variability	Treatment significance
Dry matter							
Duodenum[b]	−8.12	−13.12	−9.90	−15.90	8.31	141.33	N.S.
Ileum[b]	38.80	41.29	37.77	46.60	5.47	26.50	N.S.
Faeces[c]	68.21	70.50	66.30	75.37	1.27	3.64	$P < .05$
Cell walls							
Duodenum[b]	−4.69	6.43	−2.33	3.84	4.83	835.29	N.S.
Ileum[b]	−4.62	−3.56	−6.32	39.42	10.82	332.93	N.S.
Faeces[c]	31.98	31.24	21.61	49.81	3.29	19.53	$P < .01$
Cellulose							
Duodenum[b]	−2.93	8.65	−4.46	9.66	6.63	485.88	N.S.
Ileum[b]	−8.80	−8.24	−6.67	32.77	12.15	1070.76	N.S.
Faeces[c]	37.88	32 87	20.80	47.75	3.06	17.57	$P < .01$
Hemicellulose							
Duodenum[b]	17.19	10.11	0.93	19.85	6.74	112.21	N.S.
Ileum[b]	10.48	4.24	4.99	47.47	8.60	102.41	$P < .05$
Faeces[c]	33.41	32.05	25.32	54.28	1.92	10.61	$P < .01$

a Each coefficient is a mean of four values
b Determined by Cr_2O_3 indicator
c Determined on total collection basis

REFERENCES

Alexander, F. & Davies, M.E. (1963) Production and fermentation of lactate by bacteria in the alimentary canal of the horse and pig. *J. Comp. Path., 73,* 1–8.

Alpers, D.H. & Solin, M. (1970) The characterization of rat intestinal amylase. *Gastroenterology, 58,* 833–842.

Alpers, D.H. & Tedesco, F.J. (1975) The possible role of pancreatic proteases in the turnover of intestinal brush border proteins. *Biochim. biophys. Acta, 401,* 28–40.

Anderson, D.M. (1974) The effect of feeding on the concentration of glucose and insulin in the portal and atrial plasma in pigs. *J. agric. Sci., Camb., 82,* 29–36.

Argenzio, R. A. & Southworth, M. (1974) Sites of organic acid production and absorption in gastrointestinal tract of the pig. *Am. J. Physiol., 228,* 454–460.

Arkhipovets, A.I. (1962) Rol' amiloliticheskoi aktivnosti slyuny svinei v pishchevaritel'nom Protsesse. *Vestnik sel'khoz. Nauk., 7,* 86–88.

Arvanitakis, C. & Olsen, W.A. (1974) Intestinal disaccharidases in chronic pancreatitis. *Am. J. dig. Dis., 19,* 417–421.

Aumaitre, A., Rérat, A., Vaissade, P. & Vaugelade, P. (1973) Etude expérimentale qualitative de l'absorption intestinale des glucides après ingestion d'un repas a bas de glucose ou d'amidon. *Annls Biol. anim. bioch. biophys., 13,* 784–788.

Auricchio, S. & Ciccimarra, F. (1970) Glucoamylolytic digestion of starch in human intestinal mucosa. 7th Intern. Congr. clin. Chem. Geneva/Evian 1969 vol. 4; Digestion and intestinal absorption. pp 45–50, Karger, Basel.

Auricchio, S., Ciccimara, F., Moauro, L., Rey, F., Jos, J. & Rey, J. (1972) Intraluminal and mucosal starch digestion in congenital deficiency of intestinal sucrase and isomaltase activities. *Pediat. Res., 6,* 832–839.

Auricchio, S., Ciccimara, F., Starace, E., Vegnente, A., Giliberti, P. & Provenzale, L. (1971) Studies on intestinal digestion of starch in man. IV Glucoamylase activity of human intestinal mucosa. *Rendiconti romani Gastroenterol., 3,* 1–8.

Bailey, C.B., Kitts, W.D. & Wood, A.J. (1956) The development of the digestive enzyme system of the pig during its pre-weaning phase of growth. B. Intestinal lactase, sucrase and maltase. *Can. J. Agric. Sci., 36,* 51–58.

Banks, W., Greenwood, C.T. & Khan, K.M. (1970) The initial stages of the action on amylose of the alpha-amylase from B. subtilis, human saliva, malted rye and porcine pancreas. *Carbohyd. Res. 12,* 79–87.

Barcroft, J., McAnally, R.A. & Phillipson, A.T. (1944) Absorption of volatile acids from the alimentary tract of the sheep and other animals. *J. exp. Biol., 20,* 120–129.

Barnett, J.E.G., Ralph, A. & Munday, K.A. (1970) Structural requirements for active intestinal transport. The nature of the carrier-sugar bonding at C-2 and the ring oxygen of the sugar. *Biochem. J., 118,* 843–850.

Becker, D.E. & Terrill, D.E. (1954) Various carbohydrates in a semipurified diet for the growing pig. *Archs. Biochem. Biophys., 50,* 399–403.

Becker, D.E., Terrill, S.W., Jensen, A.H. & Hanson, L.J. (1957) High levels of dried whey powder in the diet of swine. *J. Anim. Sci., 16,* 404–412.

Becker, D.E., Ullrey, D.E. & Terrill, S.W. (1954) A comparison of carbohydrates in a synthetic milk diet for the baby pig. *Archs Biochem. Biophys., 48,* 178–183.

Becker, D.E., Ullrey, D.E., Terrill, S.W. & Notzold, R.A. (1954) Failure of the newborn pig to utilize dietary sucrose. *Science, 120,* 345–346.

Bernfeld, P., Staub, A. & Fischer, E.H. (1948) Sur les enzymes amylolytiques. XI. Propriétés de l'α-amylase de salive humaine crystallisée. *Helv. chim. Acta, 31,* 2165–2172.

Bines, B.J. & Whelan, W.J. (1960) The mechanism of carbohydrase action. 6. Structure of a salivary amylase limit dextrin from amylopectin. *Biochem. J., 76,* 253–257.

Blanco, V., Raun, N.S. & Vargas, E. (1964) Molasses as a major energy source for swine. *J. Anim. Sci., 23,* 868.

Braude, R., Dollar, A.M., Mitchell, K.G. & Porter, J.W.G. (1958) The utilization of raw and cooked starch by the young pig. *Proc. Nutr. Soc., 17,* xv–xvi.

Brooks, C.C. & Iwanaga, I.I. (1967) Use of cane molasses in swine diets. *J. Anim. Sci., 26,* 741–745.

Brown, H.T. & Heron, J. (1880) Ueber die hydrolytischen Wirkungen des Pankreas und des Dünndarms. *Justus Liebigs Annln Chem., 204,* 228–251.

Burgess, E.A., Levin, B., Mahalanabis, D. & Tonge, R.E. (1964) Hereditary sucrose intolerance: Levels of sucrase in jejunal mucosa. *Arch. Dis. Childh., 39,* 431–443.

Caldwell, M.L., Adams, M., Kung, J.T. & Toralballa, G.C. (1952) Crystalline pancreatic amylase. II. Improved method for its preparation from hog pancreas gland and additional studies of its properties. *J. Am. chem. Soc., 74,* 4033–4035.
Clemens, E.T. Stevens, C.E. & Southworth, M. (1975) Sites of organic acid production and pattern of digesta movement in the gastrointestinal tract of swine. *J. Nutr., 105,* 759–768.
Conklin, K.A., Yamashiro, K.M. & Gray, G.M. (1975) Human intestinal sucrase-isomaltase. Identification of free sucrase and isomaltase and cleavage of the hybrid into active distinct subunits. *J. biol. Chem., 250,* 5735–5741.
Cori, C.F. (1925) The fate of sugar in the animal body. I. The rate of absorption of hexoses and pentoses from the intestinal tract. *J. biol. Chem., 66,* 691–715.
Crane, R.K. (1960) Intestinal absorption of sugars. *Physiol. Rev., 40,* 789–825.
Crane, R.K. (1968) The digestive-absorptive surface of the small bowel mucosa. *Ann. Rev. Med., 19,* 57–68.
Cranwell, P.D. (1968) Microbial fermentation in the alimentary tract of the pig. *Nutr. Abs. & Rev., 38,* 721–730.
Cranwell, P.D., Noakes, D.E. & Hill, K.J. (1968) Observations on the stomach contents of suckling pigs. *Proc. Nutr. Soc., 27,* 26A.
Cranwell, P.D., Noakes, D.E. & Hill, K.J. (1976) Gastric secretion and fermentation in the suckling pig. *Br. J. Nutr., 36,* 71–86.
Cunningham, H.M. (1959) Digestion of starch and some of its degradation products by newborn pigs. *J. Anim. Sci., 18,* 964–975.
Cunningham, H.M. & Brisson, G.J. (1957) The utilization of maltose by newborn pigs. *J. Anim. Sci., 16,* 574–577.
Cunningham, H.M., Friend, D.W. & Nicholson, J.W.G. (1963) Observations on digestion in the pig using a re-entrant intestinal fistula. *Can. J. Anim., 43,* 215–225.
Dahlqvist, A. (1958) Characterisation of intestinal invertase as a glucosido-invertase. I. Action on different substrates. *Acta chem. scand., 12,* 2012–2020.
Dahlqvist, A. (1959a) Studies on the heat inactivation of intestinal invertase, maltase and trehalase. *Acta chem. scand., 13,* 945–953.
Dahlqvist, A. (1959b) Characterization of intestinal invertase as a glucoside-invertase. II. Studies on transglycosylation by intestinal invertase. *Acta chem. scand., 13,* 1659–1667.
Dahlqvist, A. (1960a) Characterisation of hog intestinal trehalase. *Acta chem. Scand., 14,* 9–16.
Dahlqvist, A. (1960b) Characterisation of three different hog intestinal maltases. *Acta chem. scand., 14,* 1–8.
Dahlqvist, A. (1960c) Characterisation of hog intestinal invertase as a glucosido-invertase. III. Specificity of purified invertase. *Acta chem. scand., 14,* 63–71.
Dahlqvist, A. (1960d) Hog isomaltase activity. *Acta chem. scand., 14,* 72–80.
Dahlqvist, A. (1960e) Substrate inhibition of intestinal glycosidases. *Acta chem. scand., 14,* 1797–1808.
Dahlqvist, A. (1960f) Hog intestinal α-glucosidases. solubilisation, separation and characterisation. Diss., Lund.
Dahlqvist, A. (1961a) Pig intestinal β-glucosidase activities. I. Relation to β-galactosidase (lactase). *Biochim. biophys. Acta 50.,* 55–61.
Dahlqvist, A. (1961b) The location of carbohydrases in the digestive tract of the pig. *Biochem. J., 78,* 282–288.
Dahlqvist, A. (1961c) Intestinal carbohydrases of a newborn pig. *Nature, Lond., 190,* 31–32.
Dahlqvist, A. (1961d) Hydrolysis of palatinose (isomaltulose) by pig intestinal glycosidases. *Acta chem. scand., 15,* 808–816.
Dahlqvist, A. (1962) Specificity of the human intestinal disaccharidases and implications for hereditary disaccharide intolerance. *J. clin. Invest., 41,* 463–470.
Dahlqvist, A. (1963) Rat-intestinal dextranase. Localization and relation to other carbohydrases of the digestive tract. *Biochem. J. 86,* 72–76.
Dahlqvist, A. & Thomson, D.L. (1963) Separation and characterization of two rat intestinal amylases. *Biochem. J., 89,* 272–277.
Decuypere, J.A. & Van der Heyde, (1972) Study of the gastro-intestinal microflora of suckling piglets and early weaned piglets using different feeding systems. *Zbl. Bakt. Hyg. I Abt. Orig., A., 221,* 492–510.
Decuypere, J.A., Vervaeke, I.J. & Hendrickx, H.K. (1974) Normal microflora of the pig's gastrointestinal tract: quantitative and qualitative composition and bacterial metabolites. 3rd Intern. Congr. (IPVS) Lyon. D31, 1–9.

Dollar, A.M., Mitchell, K.G. & Porter, J.W.G. (1957) The utilization of carbohydrates in the young pig. *Proc. Nutr. Soc., 16,* xii.
Dunkin, A.C. (1959) The effects of feeding wheys of different ages to fattening pigs. *N.Z. J. agric. Res., 2,* 1111–1120.
Ekstrom, K.E., Benevenga, N.J. & Grummer, R.H. (1975a) The effect of diets containing dried whey on the lactase activity of the small intestinal mucosa and the contents of the small intestine and caecum of the pig. *J. Nutr., 105,* 851–860.
Ekstrom, K.E., Benevenga, N.J. & Grummer, R.H. (1975b) Changes in the intestinal lactase activity in the small intestine of two breeds of swine from birth to 6 weeks of age. *J. Nutr., 105,* 1032–1038.
Ekstrom, K.E., Benevenga, N.J. & Grummer, R.H. (1975c) Effects of various dietary levels of dried whey on the performance of growing pigs. *J. Nutr., 105,* 846–850.
Ekstrom, K.E., Grummer, R.H. & Benevenga, N.J. (1976) Effects of a diet containing 40% dried whey on the performance and lactase activities in the small intestine and cecum of Hampshire and Chester White pigs. *J. Anim. Sci., 42,* 106–113.
Elsden, S.R., Hitchcock, M.W.S., Marshall, R.A. & Phillipson, A.T. (1946) Volatile acid in the digesta of ruminants and other animals. *J. exp. Biol., 22,* 191–202.
Étienne, M. (1971) Production of short chain fatty acids during digestion of the pig. *Annls Biol. anim. Biochim. Biophys., 11,* 341–342.
Farrell, D.J. (1973) Digestibility by pigs of the major chemical components of diets high in plant cell-wall constituents. *Anim. Prod., 16,* 43–47.
Ferguson, D.R. & Burton, K.A. (1977) Reconstitution in phospholipid vesicles of a glucose transport system from pig small intestine. *Nature, Lond., 265,* 639–642.
Friend, D.W., Cunningham, H.M. & Nicholson, J.W.G. (1963a) Volatile fatty acids and lactic acid in sections of the alimentary tract of the young pig. *Can. J. Anim. Sci., 43,* 174–181.
Friend, D.W., Cunningham, H.M. & Nicholson, J.W.G. (1963b) Production of organic acids in the pig. II. The effect of diet on the levels of volatile fatty acids and lactic acid in sections of the alimentary tract. *Can. J. Anim. Sci., 43,* 156–168.
Furuya, S. & Takahashi, S. (1975) Absorption of L-histidine and glucose from the jejunum segment of the pig and its diurnal fluctuation. *Br. J. Nutr., 34,* 267–277.
Gracey, M., Burke, V. & Oshin, A. (1972) Active intestinal transport of D-fructose. *Biochim. biophys. Acta, 266,* 397–406.
Gray, G.M. (1970) Carbohydrate digestion and absorption: *Gastroenterology, 58,* 96–107.
Ginsburg, V. & Hers, H.G. (1960) On the conversion of fructose to glucose by guinea-pig intestine. *Biochim. biophys. Acta, 38,* 427–434.
Hartman, P.A., Hays, V.W., Baker, R.O., Neagle, L.H. & Catron, D.V. (1961) Digestive enzyme development in the young pig. *J. Anim. Sci., 20,* 114–123.
Heyworth, R. & Dahlqvist, A. (1962) Pig intestinal β-glucosidase activities. II. Evidence for the hydrolysis of 4-methyl umbelliferyl β-D-glucoside and β-D-galactoside at the same enzyme site. *Biochim. biophys. Acta, 64,* 182–184.
Hill, K.H., Noakes, D.E. & Lowe, R.A. (1970) Gastric digestive physiology in the calf and piglet *in* Physiology of Digestion and Metabolism in the Ruminant. pp 166–179. Oriel Press, Newcastle on Tyne. pp
Holmes, J.H.G., Bayley, H.S. & Horney, F.D. (1973) Digestion and absorption of dry and high moisture diets in the small and large intestine of the pig. *Br. J. Nutr., 30,* 401–410.
Horszczaruk, F. (1962) Wplyw zróznicowanego poziomu wlókna surowego w dawkach na procesy trawienne u świń. III Rozklad wlókna surowego w jelicie ślepym. *Rocz. Nauk Roln., 80–B–3,* 333–342.
Horszczaruk, F. & Śljivovaćki, K. (1966) Rozklad wlokna surowego w jelicie ślepym i okreznicy u świń. *Roczn. Nauk roln., 89–B–1,* 67–71.
Horszczaruk, F. & Śljivovaćki, K. (1971) Rozklad wlokna surowego w jelicie ślepym i okreznicy u świń. II. *Roczn. Nauk roln., 93–B–1,* 143–147.
Iezuitova, N.N., de Laey, P. & Ugolev, A.M. (1964) Digestion of starch in vivo and in vitro in a rat intestine. *Biochim. biophys. Acta, 86,* 205–210.
Johnson, S.R. (1949) Comparison of sugars in the purified diet of baby pigs. *Fedn. Proc. Fedn. Am. Socs. exp. Biol., 8,* 387.
Keys, J.E. & DeBarthe, J.V. (1974a) Site and extent of carbohydrate, dry matter, energy and protein digestion and the rate of passage of grain diets in swine. *J. Anim. Sci., 39,* 57–67.
Keys, J.E. & DeBarthe, J.V. (1974b) Cellulose and hemicellulose digestibility in the stomach, small intestine and large intestine of swine. *J. Anim. Sci., 39,* 53–56.

Kidder, D.E., Hill, F.W.G. & Stevens, J.A. (1972) The automatic measurement of some mucosal carbohydrases. *Clin. Chim. Acta, 38,* 491–501.

Kidder, D.E. & Manners, M.J. (1975) Carbohydrases in the small intestine mucosa of the pig. *Proc. Nutr. Soc., 35,* 27A–28A.

Kidder, D.E., Manners, M.J. & McCrea, M.R. (1963) The digestion of sucrose by the piglet. *Res. vet. Sci., 4,* 131–144.

Kidder, D.E., Manners, M.J., McCrea, M.R. & Osborne, A.D. (1968) Absorption of sugars by the piglet. *Br. J. Nutr., 22,* 501–514.

Kim, K.I., Benevenga, N.J. & Grummer, R.H. (1976) Lactose fermentation in the caecum and colon of pig. *J. Anim. Sci., 42,* 1350.

Kolinská, J. & Semenza, G. (1967) Studies on intestinal sucrase and on intestinal sugar transport. 5. Isolation and properties of sucrase-isomaltase from rabbit small intestine. *Biochim. biophys. Acta., 146,* 181–195.

Kvasnitskii, A.V.(1951) Voprosy fiziologii pishchevareniya u svinei. *Sel'khozgiz,* Moscow.

Labat, J., Baumann, F. & Courtois, J.E. (1973) Hydrolyse du tréhalose par la tréhalase de porc: configuration anomérique α du D-glucose libéré. *Carbohydrate Research, 26,* 341–349.

Larner, J. & Gillespie, R.E. (1956) Gastrointestinal digestion of starch. II. Properties of the intestinal carbohydrases. *J. biol. Chem., 223,* 709–726.

Larner, J. & McNickle, C.M. (1955) Gastrointestinal digestion of starch. I. The action of oligo-1, 6-glucosidase on branched saccharides. *J. biol. Chem. 215,* 723–726.

Ledinek, M. (1970) Untersuchungen über die Beteiligung von Mikroorganismen am Verdauungsstoffwechsel des Schweines bei Fütterung roher und gedämpfter Kartoffeln. Diss. Munich.

Loginov, G.I. (1973) Kompensatornye reaktsii tonkoi kishki posle vyklyucheniya podzheludochnoi zhelezy iz pishchevareniya. *Sechenov Fiziol. Zh. S.S.S.R., 59,* 802–807.

Ly, J. (1971) Intestinal digestion of molasses in the growing pig. *Rev. cubana Cienc. agric. (Eng. ed.), 5,* 351–361.

Ly, J. (1975) Caecal function in the pig: daily variation of caecal parameters in pigs fed on molasses diets. *Rev. cubana Cienc. agric. (Eng. ed.), 9,* 39–47.

Ly, J. & Boucourt, R. (1975) Microflora and fermentation in the gastrointestinal tract of the young pig. 1. Levels of organic acids. *Rev. cubana agric. Cienc. (Eng. ed.), 9,* 149–162.

Ly, J. & Velásquez, M. (1970) Some observations on blood glucose level in pigs given diets based on final molasses and sugar, high test molasses or grain. *Rev. cubana Cienc. agric. (Eng. ed.), 4,* 195–199.

Maestracci, D. (1976) Enzymic solubilization of the human intestinal brush border membrane enzymes. *Biochim. biophys Acta, 433,* 469–481.

Manners, M.J. & Stevens, J.A. (1972) Changes from birth to maturity in the pattern of distribution of lactase and sucrase activity in the mucosa of the small intestine of pigs. *Br. J. Nutr., 28,* 113–127.

Mason, V.C. & Just, A. (1976) Bacterial activity in the hind-gut of pigs. 1. Its influence on the apparent digestibility of dietary energy and fat. *Z. Tierphysiol. Tierernähr. Futtermittelk., 36,* 301–310.

MacLeod, N.A., Preston, T.R., Lassota, L., Willis, M.B. & Velasquez, M. (1968) Molasses and sugar as energy sources for pigs. *Rev. cubana Cienc. agric. (Eng. ed.), 2,* 205–210.

McMichael, H.B. & Dahlqvist, A. (1968) Some studies on starch hydrolysis. An interpretation of membrane digestion. *Gut, 9,* 365–366.

Mendel, L.B. & Mitchell, P.H. (1907) Chemical studies on growth. 1. The inverting enzymes of the alimentary tract, especially in the embryo. *Am. J. Physiol., 21,* 81–96.

Messer, M. & Kerry, K.R. (1967) Intestinal digestion of maltotriose in man. *Biochim. biophys. Acta, 132,* 432–443.

Meyer, K.H. & Gonon, W.F. (1951) La degrédation de l'amidon par les α-amylases. Recherches sur l'amidon 50. *Hel. chim. Acta. 34,* 294–307.

Myrbäck, K. (1926) Über Verbindungen einiger Enzyme mit inaktivierenden Stoffen. 2. *Hoppe-Seyler's Z. physiol. Chem., 159,* 1–84.

Noakes, D.E. (1972) The development of digestive function in the baby pig. Thesis, London.

Ogata, M. & Morishita, Y. (1969) Studies on the alimentary flora of pigs. IV. The alimentary flora of pigs infected with hog cholera. *Jap. J. vet. Sci., 31,* 71–82.

Plimmer, R.H.A. (1907) On the presence of lactase in the intestines of animals and on the adaptation of the intestine to lactose. *J. Physiol., Lond., 35,* 20–31.

Preiser, H., Menard, D., Crane, R.K. & Cerda, J.J. (1974) Deletion of enzyme protein from the brush border membrane in sucrase-isomaltase deficiency. *Biochim. biophys. Acta, 363,* 279–282.

Ramaswamy, K., Malathi, P., Caspary, W.F. & Crane, R.K. (1974) Studies on the transport of glucose from disaccharides by hamster small intestine in vitro. II Characteristics of the disaccharidase-related transport systems. *Biochim. biophys. Acta, 345,* 39–48.
Rensmann, F.W. (1976) Zur Bedeutung wechselnder Rohfaser- und Stärkemengen im Futter für den Gehalt Trockensubstanz, flüchtigen Fettsäuren, Milchsäure und Ammoniak im Kot von Ferkeln. Diss., Hannover.
Rérat, A., Aumaitre, A., Vaugelade, P. & Vaissade, P. (1974) Quantitative variations in the absorption of glucose during digestion of maize starch in the pig. *Proc. Nutr. Soc., 33,* 102A–103A.
Rérat, A., Février, C., Aumaitre, A., Vaissade, P. & Vaugelade, P. (1973) Absorption intestinale de quelques glucides chez le porc. Etude expérimentale qualitative par la mesure en continu de la glycémie porte et jugulaire. *Cah. Nutr. Diet., 8,* 154–156.
Roberts, P.J.P. & Whelan, W.J. (1960) The mechanism of carbohydrase action. 5. Action of human salivary α-amylase on amylopectin and glycogen. *Biochem. J. 76,* 246–253.
Robyt, J.F. & French, D. (1967) Multiple attack hypothesis of α-amylase action; action of porcine pancreatic, human salivary and Aspergillus oryzae α-amylases. *Archs Biochem., 122,* 8–16.
Robyt, J.F. & French, D. (1970a) The action pattern of porcine pancreatic α-amylase in relationship to the substrate binding site of the enzyme. *J. biol. Chem., 245,* 3917–3927.
Robyt, J.F. & French, D. (1970b) Multiple attack and polarity of action of porcine pancreatic α-amylase. *Archs. Biochem., 138,* 662–670.
Ruttloff, H., Friese, R. & Täufel, K. (1964) Zur enzymatischen Spaltung von Leucrose. *Naturwissenschaften, 51,* 163.
Ruttloff, H., Friese, R. & Täufel, K. (1967a) Die Bedeutung glucosidatischer Enzyme bei der Zerlegung von Maltooligosacchariden und Stärke. *Hoppe-Seyler's Z. physiol. Chem., 348,* 705–710.
Ruttloff, H., Friese, R. & Täufel, K. (1967b) Zur Bestimmung der intestinalen γ-Amylase. *Nahrung, 11,* 205–213.
Ruttloff, H., Friese, R. & Täufel, K. (1968) Zur Bedeutung der α-amylase bei der intestinalen Stärkehydrolyse. *Ernährungsforschung, 13,* 141–154.
Sato, M. & Yamashina, I. (1974) Fractionation and characterization of β-galactosidases from hog small intestine. *J. Biochem., Tokyo, 70,* 683–693.
Semenza, G. (1968) Digestion and absorption of sugars. *Mod. Probl. Pediat., 11,* 32–47.
Semenza, G., Curtius, C. -H., Kolinská, J. & Müller, M. (1967) Studies on intestinal sucrase and on intestinal sugar transport. VI Liberation of α-glucose by sucrase and isomaltase from glycone moiety of substrates. *Biochim. biophys. Acta, 146,* 196–204.
Shearer, I.J. & Dunkin, A.C. (1968a) Lactose utilization by the growing pig. *N.Z. J. agric. Res., 11,* 465–476.
Shearer, I.J. & Dunkin, A.C. (1968b) Caecum enlargement in lactose-fed pigs. *N.Z. J. agric. Res., 11,* 923–926.
Sewell, R.F. & Maxwell, C.V. (1966) Effects of various sources of carbohydrates in the diet of early weaned pigs. *J. Anim. Sci., 25,* 796–799.
Slivitskii, M.G. (1973) Utvorennya molochnoi kisloti v travnomu trakti svinei. *Svinovodstvo, Kiev, 26,* 74–80.
Slivitskii, M.G. (1975) Brodil'nye protsessy v zheludke svinei. *Vest. sel'khoz. Nauk., 7,* 75–80.
Smith, H.W. & Jones, J.E.T. (1963) Observations on the alimentary tract and its bacterial flora in healthy and diseased pigs. *J. Path. Bact., 86,* 387–412.
Stevens, J.A. & Kidder, D.E. (1972) The distribution of trehalase, sucrase, glucoamylase and lactase (β-galactosidase) along the small intestine of five pigs. *Br. J. Nutr., 28,* 129–137.
Tacu, A. & Bianu, G. (1974) Maltaza intestinala si pancreatica la porcine. Lucrarile stiintifice ale Institutului de Cercetari pentru. *Nutritia animalelor, 3,* 155–176.
Täufel, K., Krause, W., Ruttloff, H. & Maune, R. (1967) Zur intestinalen Spaltung von Oligosacchariden. *Z. ges. exp. Med., 144,* 54–66.
Ugolev, A.M. (1960) On the existence of membrane (contact) digestion. *Biul. Eksp. Biol. Med., 49,* 12–17.
Van der Heyde, H. & Hendrickx, H. (1964) Quantitative Untersuchungen der Flora des Verdauungskanals von Saugferkeln. *Zbl. Bakt. Parasitenk. Hyg. Org., 195,* 215–226.
Velasquez, M., Ly, J. & Preston, T.R. (1969) Digestible and metabolizable energy values for pigs of diets based on high-test molasses or final molasses and sugar. *J. Anim. Sci., 29,* 578–580.

Walker, D.M. (1959) Development of the digestive system of the young animal. II. Carbohydrase enzyme development in the young pig. *J. Agric. Sci., Camb., 52,* 357–363.
Walker, G.J. & Whelan, W.J. (1960) The mechanism of carbohydrase action. 7. Stages in the salivary α amylolysis of amylose, amylopectin and glycogen. *Biochem. J., 86,* 257–263.
Wallenfels, K. & Bernt, E. (1953) Uber den Verlauf der enzymatischen Spaltung von Lactose mit β-Galactosidase von Schimmelpilzen, Helix pomatia, Escherichia coli und Kälberdarm. *Justus Liebigs Annln. Chem., 584,* 63–85.
Whelan, W.J. & Roberts, P.J.P. (1953) The mechanism of carbohydrase action. 2. α-amylolysis of linear substrates. *J. chem. Soc., 1953,* 1298–1304.
Widdowson, E.M. Colombo, V.E. & Artavanis, C.A. (1976) Changes in the organs of pigs in response to feeding for the first 24h after birth. II. The digestive tract. *Biol. Neonate, 28,* 272–281.
Wilson, T.H. (1962) Intestinal absorption, pp 69–109. W.B. Saunders, Philadelphia.
Wise, M.B., Barrick, E.R., Wise, G.H. & Osborne, J.C. (1954) Effect of substituting xylose for glucose in a purified diet for pigs. *J. Anim. Sci., 13,* 365–374.

CHAPTER 7
DIGESTION OF PROTEIN

SUMMARY

The proteins in the food are first acted on in the stomach by the pepsins produced from the pepsinogens of the gastric secretions. The pepsins act in the very acid conditions of the fundal and pyloric regions of the stomach and start the breakdown of proteins to large peptides.

When the contents of the stomach pass into the small intestine, the pH is raised by the pancreatic juice, the bile and the secretions of the intestine wall, producing the conditions for the proteins and peptides to be acted on by the pancreatic enzymes. Of these, the trypsin, chymotrypsin and elastase continue the breakdown to give progressively smaller peptides while the carboxypeptidases release amino acids from the carboxy end of suitable peptides.

The aminopeptidases on the mucosal surface hydrolyse peptides to amino acids which are then actively absorbed by the mucosal surface. The mucosal surface also absorbs some small peptides, which are hydrolysed intracellularly.

GASTRIC DIGESTION

Digestion of protein starts in the stomach as a result of the action of the pepsins of the gastric juice. Like other proteolytic enzymes throughout the body, they are formed as inactive precursors which are activated by hydrolytic removal of a peptide from the N-terminal end of the molecule. The precursors of the pepsins are called pepsinogens, and are secreted by the mucosa of the fundal and pyloric regions of the stomach.

Pig pepsinogens are hydrolysed to pepsins in acid conditions, slowly at pH 4 and rapidly at pH 2 (Ryle, 1960), the pepsin produced catalysing the activation, so that the process is autocatalytic. At pH 4 the intramolecular activation due to the acidity alone is slow, so that activation is predominantly by the pepsin, but below pH 3 the intramolecular reaction is rapid and predominant (Al-Janabi, Hartsuck & Tang, 1972). The intramolecular activation involves a split between amino acid residues 44 and 45 (Sanny, Hartsuck & Tang, 1975) while the autocatalytic activation is due to hydrolysis of a glutamyl-isoleucyl bond with the release of a 41 amino acid peptide which then can act as an inhibitor of pepsin but is hydrolysed when pepsin is present in excess. This is probably a protective device which inhibits premature activation of the enzyme within the secretory gland, but permits rapid activation after secretion.

The pepsins each have two pH optima, one near pH 2.0, and the other about 3.5 (Taylor, 1959a), the pepsins from the pyloric region having slightly lower pH optima than those from the fundal region (Table 7.1). This is consistent with the fact that the pH in the pyloric region is normally lower than that in the fundus. The way in which activity falls off with increasing pH above 3.5 depends on the substrate (Taylor, 1959a), but pig pepsin has not been found to show any activity above pH 6.0 (Taylor, 1959b).

Table 7.1 *Pig pepsins*

Name	Other names	Origin	Lower	Higher	References
Pepsin A	Pepsin	Fundal mucosa	2.0–2.2	3.5–3.9	Taylor (1962)
Pepsin D	–				Lee & Ryle (1967a, b)
Pepsin B	Parapepsin I	Pyloric mucosa	1.7–1.9	3.2–3.4	Ryle & Porter (1959)
Pepsin C	Parapepsin II				Ryle (1960)

A summary of the specificity of pepsin A based on the experiments of a variety of authors using synthetic peptides, usually dipeptides, is given by Bovey & Yanari (1960). Similar findings have been obtained with larger peptides of known composition from natural sources (Etherington & Taylor, 1971; Carnegie, 1971; Haylett, Swart & Parris, 1971; Shotton & Hartley, 1973). The main conclusions are as follows. Peptide bonds are only attacked if both the contributory amino acids are of the naturally-occurring L-configuration. Aromatic substituted amino acids (e.g. tyrosine and phenylalanine) contributing to the peptide bond produce most rapid hydrolysis, followed by glutamic acid, and then cysteine or cystine. Glycyl peptides have very low susceptibility to pepsin. Action is only rapid if the amino acid on at least one side of the bond has an aromatic or other suitable side chain, and most rapid if both the amino acids are of this type (Inouye & Fruton, 1967).

The specificities of the individual pepsins differ slightly, according to Etherington & Taylor (1971), who state that the specificities for any one pepsin are the same at the two pH optima. On the other hand Taylor (1959a) found that the pepsins of pig and man, but not the calf, only showed the lower pH optimum with egg albumen, but showed both with a number of other proteins. He also reported that peptides of aromatic amino acids were more rapidly attacked at the more acid pH, whereas the glutamyl peptides were more susceptible at the less acid optimum (Taylor, 1962).

Kratzer & Porter (1962) showed that the ratio of activity of pig pepsin at the two maxima was different for different proteins, and that the peptide mixtures produced from proteins were not the same at the two maxima. It is not known whether this is due to the effect of pH on the enzyme or the substrate.

The pH of the contents in the fundal and pyloric regions of the pig's stomach falls rapidly after a meal to below 5.0 and then progressively to around 2.0 (Lawrence, 1972) so that, first slow, and then more rapid peptic hydrolysis of the proteins can occur.

A number of early in vitro studies suggested that once a protein molecule had been partially hydrolysed by pepsin, it was very rapidly further degraded to small peptides, so that a partial hydrolysate consisted largely of intact protein and small peptides (Tiselius & Eriksson-Quensel, 1939; Moring-Claesson, 1948; Desnuelle, Rovery & Bonjour, 1950). Gel chromatography techniques have shown that this is not the case, and Steinhart & Kirchgessner (1973) found that soya protein, when digested by pig pepsin, was progressively reduced through large and then medium, to small peptides, so that first the large, then the intermediate and finally the small peptides predominated in the hydrolysate. Beyer, Steinhart & Kirchgessner (1975) found that while alteration in the enzyme: substrate ratio affected the extent to which this breakdown to smaller peptides occurred, it did not have a consistent influence on the amino acid composition of the products.

The rate of digestion varies considerably from one type of protein to another. Nehring & Schröder (1965) compared the peptic digestion of the proteins of lucerne, rape, barley and lupin with meat flour. Amino acids and small peptides were released in the greatest quantity from the rape. The amino acids preferentially released either as free amino acid or as small peptides also varied from one protein to another. No detailed studies of the course of protein breakdown have been reported using actual digesta from a pig.

Kirchgessner, Beyer & Steinhart (1976) found that copper had a marked activating effect on peptic digestion, and suggested that this contributed to the improved performance of pigs on copper supplements.

DIGESTION IN THE SMALL INTESTINE

As the contents of the stomach pass through the duodenum they are mixed with pancreatic and duodenal secretions, and bile. All these secretions are alkaline, with the result that the pH rises progressively, as described in Chapter 3, reaching nearly 7 by the end of the small intestine. This takes the digesta out of the pH range of the pepsins and into the range of the pancreatic proteolytic enzymes. The amino acids formed by the protein breakdown are absorbed, and the peptides further broken down by the intestinal mucosal enzymes before or after absorption.

The proteolytic enzymes involved in intestinal digestion can be classified in three groups (Table 7.2). These are:—

1. Endopeptidases which act on a susceptible peptide link wherever it occurs in the protein chain, provided it is accessible to the enzyme.
2. Carboxypeptidases, which remove an amino acid residue from the carboxyl end of the chain.
3. Aminopeptidases, which remove an amino acid residue from the amino end of the chain.

The endopeptidases and carboxypeptidases are secreted by the pancreas as inactive precursors called "zymogens", while the amino-peptidases are associated with the intestinal mucosa, being present either on the brush border or within the mucosal cell. The zymogens of all the pancreatic proteases are activated by trypsin which splits an arginyl or lysyl bond to release a small peptide from the N-terminal end of the molecule, the rest of the molecule forming the active enzyme. Initially trypsinogen is activated by enterokinase (enteropeptidase), an enzyme located on the brush border of the duodenal mucosa. The trypsin so produced can activate more trypsinogen as well as activating the other proenzymes.

While the levels of all the enzymes in the pancreatic juice rise in response to CCK stimulation of the pancreas due to the presence of food in the duodenum, a specific rise in the chymotrypsin content occurs when the pig is on a high protein diet (Corring & Saucier, 1972; Chapter 5).

Table 7.2 Proteases present in the intestines

Class	Enzyme	Zymogen	Origin	Site of action	Approximate concentration in pig pancreatic juice as % of protein (Marchis-Mouren) 1959	1965
Endopeptidase	Trypsin	Trypsinogen			12	23
	Chymotrypsin A	Chymotrypsinogen A				
	Chymotrypsin B	Chymotrypsinogen B				
	Chymotrypsin C	Chymotrypsinogen C			5	10.5
	Elastase	Proelastase	Pancreas	Small intestine lumen		
Exopeptidase	Carboxy-peptidase A	Procarboxy-peptidase A			18	22
	Carboxy-peptidase B	Procarboxy-peptidase B			16	16
	Amino-peptidase		Small intestine mucosa	Surface of small intestine mucosa		

Pancreatic Enzymes

The enzymes and inhibitors found in pancreatic juice are listed in detail in Tables 5.4, 5.5 & 5.6. Their digestive function will be described here.

Endopeptidases

The pancreatic endopeptidases will hydrolyse the amides and esters of suitable amino acids, whether these occur free or at any accessible position in a protein or peptide chain. The enzymes are of three types depending on the amino acid contributing the carboxyl portion of the bond. If this amino acid has a basic side chain (e.g. lysine), the bond will be susceptible to trypsin, whereas chymotrypsin requires a substantial but non-basic side-chain (e.g. tyrosine) and elastase requires the side-chain to be short or absent (e.g. alanine). Elastase II is less specific, and has the properties of both a chymotrypsin and an elastase.

TRYPSIN

Trypsins, which are secreted as inactive precursors, the trypsinogens, are the most important pancreatic endopeptidases, not merely because they are present in the greatest quantity, but because they are the activators of the precursors of the other pancreatic proteolytic enzymes and also of pancreatic phospholipase. The activation of trypsinogen is thus the key process in the activation of pancreatic juice. This activation is affected by the hydrolysis of a lysine-isoleucine bond with the removal of a peptide from the N-terminal end of trypsinogen (Charles, Rovery, Guidoni & Desnuelle 1963) initiated by the action of enterokinase, an enzyme attached to the duodenal brush border (Holmes & Lobley, 1970; Nordström & Dahlqvist, 1970, 1971; Hadorn, Steiner, Sumida & Peters, 1971), and released from it by trypsin (Barns & Elmslie, 1977).

The activation continues as an autocatalytic reaction with the trypsin activating more trypsinogen. The relevant lysine-isoleucine bond is rendered more resistant to tryptic hydrolysis by the sequence of four aspartyl residues which immediately precede the lysine in the trypsinogen of the pig and some other species (Abita, Delaage, Lazdunski & Savdra, 1969). This will reduce the likelihood of premature activation of the enzymes in the pancreas and pancreatic duct. In addition, trypsin inhibitors are present in pancreatic juice (Kunitz & Northrop, 1936). These are peptides which are destroyed by hydrolysis if trypsin is present in excess, so that the activation process is only inhibited until sufficient activator is present, after which it becomes self-accelerating (Fritz, Hüller, Wiedemann & Werle, 1967). These inhibitors have no effect on activation of trypsin by enterokinase (Maroux, Baratti & Desnuelle, 1971), and so will not interfere with activation of trypsin in the duodenum.

Trypsin is the most specific of the pancreatic endopeptidases, acting only on those bonds where the carboxyl is contributed by an amino acid with a basic side chain (in natural proteins this means arginine and lysine), and having no detectable action on other bonds (Travis, 1967). Two trypsinogens occur in pig pancreatic juice, one being a cationic protein which has been known for a long time, and the other being anionic (Voytek & Gjessing, 1971), both being activated by enterokinase to corresponding trypsins with the loss of the same octapeptide, the anionic rather more readily (Louvard & Puigserver, 1974). Voytek & Gjessing (1971) reported the anionic trypsin to be more inhibited by the pancreatic trypsin inhibitor and to account for about half the trypsin activity but Louvard & Puígserver (1974) regarded it as forming less than 10% of the total trypsin.

The rate of hydrolysis of a natural protein by trypsin varies very much from one protein to another. Nehring & Schröder (1965) compared in vitro trypsin digestion of various vegetable proteins with meat flour, and found that release of amino acids and small peptides was very much greater with the meat flour than with any of the vegetable proteins. Among the latter, protein from rape was much less digested than that from lucerne, barley or lupin.

CHYMOTRYPSINS

The chymotrypsins are less specific than trypsin. They have the most rapid action on bonds where the carboxyl is contributed by an amino acid with an aromatic side chain (tyrosyl, phenylalanyl, tryptophanyl) but also have quite rapid action on many bonds where the relevant amino acid has instead an aliphatic side chain (leucyl, methionyl, asparaginyl, glutaminyl). The chymotrypsins have no action on the bonds split by trypsin (arginyl, lysyl) (Folk & Cole, 1965).

As the other substrate amino acids in the vicinity of the bond also influence the susceptibility to hydrolysis, not all bonds of a suitable amino acid will be hydrolysed at a measurable rate by any particular chymotrypsin. The specificity of the different chymotrypsins probably differs, so that in a given protein or polypeptide, the different chymotrypsins may preferentially split different bonds and so have complementary effects.

ELASTASE

Elastase (Pancreatopeptidase E) was originally so named because it hydrolyses the connective-tissue protein elastin, which is unaffected by trypsin or chymotrypsin A (Baló & Banga, 1949, 1950). Pig pancreatic elastase acts most rapidly on alanyl bonds, but also hydrolyses glycyl, seryl, valyl, and, in some situations, leucyl bonds (Narayanan & Anwar, 1969).

ELASTASE II

This enzyme, perhaps identical with proteases from pig pancreas which have previously been reported under other names (Chapter 5, Table 5), acts both on typical chymotrypsin substrates (tyrosine esters and amides) and on elastin and other elastase substrates (Ardelt, 1974, 1975; Ledoux & Lamy, 1975).

Carboxypeptidases

The carboxypeptidases are secreted in high concentration in pig pancreatic juice in the form of precursors called procarboxypeptidases (Table 2) which are activated by tryptic removal of the N-terminal end of the molecule. They catalyse the hydrolytic removal of one amino acid from the carboxy end of a peptide. As they can only act on a chain end, they will have only very slow action on large protein molecules where there are few chain ends in relation to the number of amino acid units, but will act rapidly on a protein hydrolysate where the chain length of the peptides is short and there are many more free chain ends for a given number of amino acid units. Their specificities, which are determined mainly by the terminal amino acid at the carboxy end of the substrate, are also such that they are particularly effective on a trypsin or chymotrypsin hydrolysate. They thus continue very effectively the proteolysis carried out by the pancreatic endopeptidases.

There are two classes of carboxypeptidases which differ in their specificity. Carboxypeptidase A acts on peptides where the terminal amino acid does not contain a basic side-chain. It is particularly active where the terminal amino acid has an aromatic side-chain (Folk & Schirmer, 1963), as in the free end produced by chymotryptic hydrolysis.

Carboxypeptidase B acts most rapidly where the terminal amino acid has a basic side chain (arginine or lysine), as in the free end produced by tryptic hydrolysis. Pig carboxypeptidase B has also been shown to have some action where the terminal amino acid is phenylalanine (Sokolovsky & Zisapel, 1971). There is thus some overlap in the activity of the two types of carboxypeptidase.

Nucleases

These enzymes hydrolyse nucleic acids, breaking the chains and ultimately degrading them to individual mononucleotides.

RIBONUCLEASE

Ribonuclease activity was demonstrated in pig pancreas extracts by Jones (1912, 1920). In comparative studies, Zendzian & Barnard (1967) showed that very large species differences occurred in the levels of pancreatic ribonuclease, which were very high in ruminants and some other groups with a ruminant-like digestion, and very low in most other species, the pig having an intermediate level. Zendzian & Bernard considered that the main function of the ribonuclease was the splitting of bacterial ribonucleic acid, so that the level of this enzyme in the pig was regarded as related to the degree of bacterial fermentation in the stomach, which is intermediate between the ruminant and the non-ruminant. Failure to break down the nucleic acid could result in appreciable loss in the faeces of dietary phosphate which had been converted to nucleic aids by the bacteria.

Pig pancreatic ribonuclease has been isolated pure and its properties described by Yamasaki, Murakami, Irie & Ukita (1968). It was shown by Irie (1968) to be similar to other mammalian pancreatic ribonucleases but with slight differences in specificity.

DEOXYRIBONUCLEASE

Marchis-Mouren, Charles, Ben Abdeljlil & Desnuelle (1961), and Marchis-Mouren (1965), fractionating pig pancreatic juice on ion exchange columns, were able to identify a fraction as deoxyribonuclease.

Mucosal Digestion of Proteins

The absorption of the digestion products of proteins by the mucosal cells has been reviewed by a number of authors (e.g. Fisher, 1967; Matthews, 1971).

It involves five processes:–

1. Hydrolysis of peptides in the lumen by aminopeptidases on the luminal surface of the mucosal brush border.
2. Absorption of free amino acids from the lumen.
3. Absorption of small peptides.
4. Hydrolysis of absorbed small peptides by aminopeptidases within the mucosal cell.
5. Metabolism of some absorbed amino acids including some resynthesis of protein.

Aminopeptidases

The enzymes situated in the intestinal mucosa and which complete the protein breakdown are all aminopeptidases, that is, enzymes which remove one amino acid at a time from the nitrogen terminal end of a peptide. Some of these are located on the brush border and so are able to act on peptides still in the lumen of the gut, while other aminopeptidases are inside the mucosal cell where they will only act on peptides which have been absorbed. The number of peptidases of this type has not been determined with certainty in any species. There are clearly several with partly overlapping specificity. At present they are defined either by a particular peptide which they are known to hydrolyse, or by the nitrogen terminal amino acid of a substrate, this being the main factor determining whether it is hydrolysed.

There have been several studies of the aminopeptidases in pig small intestine mucosa. Josefsson & Lindberg (1965a) measured the activity against alanyl-glutamic acid, glycyl-glycine, glycyl-leucine and glycyl-valine. They showed that these four peptidase activities were differently affected by the presence of Co^{++}, Mn^{++} and Zn^{++}. It is therefore probable that they are due to different enzymes. They have studied the distribution of these activities along the pig small intestine (Josefsson & Lindberg, 1965b). Only glycyl-glycine hydrolase activity was found at the extreme duodenal end, but otherwise the four activities occurred along the whole small intestine, being greatest in the middle third and gradually diminishing towards each end.

Norén, Sjöström & Josefsson (1973) have isolated pure a depeptidase with high activity against glycyl-L-leucine, and Sjöström, Norén & Josefsson, (1973) and Sjöström & Norén (1974), a prolidase from pig jejunal mucosa. Norén, Dabelsteen, Sjöström & Josefsson (1977) showed that these two enzymes were located in the cytoplasm of the enterocytes.

Maroux, Louvard & Baratti (1973) and Maroux & Louvard (1977) have isolated from the brush border of pig intestinal mucosa an aminopeptidase complex with a molecular weight of 280,000 which can be broken down to three subunits of molecular weights of about 130,000, 97,000 and 49,000. This complex seems to be responsible for all the brush border aminopeptidase activity, and all the mucosal activity against arylamides and the heptapeptide used as substrate. It was responsible for half the mucosal tripeptidase activity and a small but significant part of the activity against dipeptides.

It has been suggested (Peters, 1970) that the larger peptides are split by the aminopeptidases on the brush border while the smaller ones are absorbed intact and hydrolysed by aminopeptidases within the mucosal cells, a suggestion which is supported by the studies of Kim, Kim & Sleisenger (1974) on the mucosal peptidases of rat and man.

Absorption

MUCOSAL TRANSPORT

These processes have been studied predominantly in laboratory rodents where it has been shown that amino acids are absorbed against a concentration gradient by energy-dependent processes, and that different absorption systems act on different groups of amino acids. This subject has been very fully reviewed by Matthews (1975). It appears that one system exists for glycine, proline and hydroxyproline, another for the other neutral amino acids, a third for the basic amino acids and cystine, and a fourth for the acidic amino acids. In vitro studies by Harmeyer, Martens & Birck (1974) on duodenal and jejunal mucosa from 8–12 week old pigs have shown that the absorption of amino acids in the pig is influenced by the same factors and inhibited by same inhibitors as in other mammals.

It was shown long ago (e.g. Messerli, 1913), but not universally accepted, that polypeptide mixtures were absorbed more rapidly than the constituent amino acids. More recent studies comparing the uptake of specific peptides with that of their constituent amino acids (Matthews, Craft, Geddes, Wise & Hyde, 1968; Craft, Geddes & Matthews, 1968; Matthews, Lis, Cheng & Crampton, 1969) have made it clear that many small peptides are absorbed as such by the mucosal cells and hydrolysed intracellularly. The absorption of peptides independently of that of amino acids (Matthews, 1975) would be expected in view of the findings, already referred to, on the distribution of aminopeptidases within the mucosal cells.

A consequence of the absorption of peptides rather than individual amino acids from the lumen is that the different amino acids from the protein in the diet will tend to be absorbed together, even though, in vitro, the individual amino acids have different rates of absorption. Such an unselective entry is also suggested by findings on blood levels. It is supported by experiments by Polivoda & Gritsenko (1973), who fed a protein diet supplemented with crystalline methionine and lysine to two fistularized 4-month old pigs. The supplementary acids disappeared from the intestinal contents early, while the amino acids originating from the protein gradually disappeared later, in parallel with one another. On the other hand, Buraczewska & Zebrowska (1972) found that when casein or a casein hydrolysate was introduced into an isolated intestinal loop in a pig, some amino acids, especially glutamic, were much more slowly absorbed than others.

BLOOD AMINO ACIDS

Amino acids resulting from protein digestion pass from the mucosal cells of the small intestine into the portal blood, but there is considerable evidence (Fisher, 1967) that this is not the only way digested protein leaves the mucosal cell. Fisher suggests that a substantial fraction of the absorbed amino acids passes from the mucosal cell into the intestinal lymph in the form of protein. Nevertheless, the amino acid level of the portal blood reflects the degree of absorption and has been used to a very limited degree to study protein digestion in the pig.

Pion, Fauconneau & Rérat (1964; 1966) measured the free amino acids in the portal blood of 20–25 kg pigs after feeds of protein-free diets with different proteins. They found no rise in amino acids after the protein-free diets, but after feeding the diets containing protein there were rises in the portal blood amino acids, which were maximal after about 3 hours. The proportions of these amino acids, and particularly those of the essential amino acids, reflected the composition of the diet. Braude, Mitchell, Myres, Porter & Williams (1969), looking at systemic blood in 20–60 kg pigs, found rises in essential amino acids after feeding, in proportions related to the composition of the diet, but found little change in the non-essential amino acids.

Looking at differences between portal and systemic blood levels, Braude, Cutts, Myres & Porter (1970) and Rérat & Duée (1973) found that the individual essential amino acid levels were higher in the portal blood of pigs after a protein-containing food to an extent reflecting the composition of the food. Shimada & Zimmerman (1973) found the same thing in 1 to 3 week old piglets. Rérat, Corring, Vaissade & Vaugelade (1977) used the difference between amino acid levels in portal and systemic blood to assess the effect on protein digestion of diverting pancreatic juice. They found the effect on protein digestion was greater with high protein diets.

Clearly, the comparison of portal and peripheral blood amino acid levels is a good indicator of protein digestion, but owing to its technical difficulty, it has been little used so far. Measurement of peripheral blood amino acid levels after feeding has been used by Orok & Bowland (1975) for comparison of dietary proteins in pigs, but differences between the times required to reach maximum levels on the different diets prevented these measurements from being reliable indicators.

DIGESTION IN THE LARGE INTESTINES

In the large intestine there is considerable bacterial deamination of amino acids (Michel, 1966; Fauconneau & Michel, 1970). Zebrowska (1973a) showed that if a casein hydrolysate was infused into the lower ileum, most of the nitrogen was absorbed, but was then excreted in the urine. She found that when such a hydrolysate was infused into the terminal ileum, about 20% of the nitrogen of the hydrolysate was retained, and the remainder excreted as urinary urea and so will have been useless to the pig (Zebrowska, 1975).

Delort-Laval, Charlet-Lery & Zelter (1963) showed that giving oral chlortetracycline to pigs reduced the proportion of absorbed dietary nitrogen which was excreted in the urine. Michel, Jouandet, Salmon-Legagneur, Aumaitre & Francois (1954) reported a reduction in faecal ammonia in pigs given oral sodium acrylate, and considered that there was a correspondence between the growth stimulating effect of an antibiotic and its effect in suppressing bacterial deamination in the large intestines. Chachulowa (1964) found a marked reduction in caecal ammonia when cellulose was included in the diet, and a smaller reduction when copper sulphate or chlortetracycline was included. Rensmann (1976) found that if wood pulp or potato starch was added to a low fibre diet fed to growing pigs, the faecal ammonia was reduced. Presumably there was either less deamination when alternative sources of energy were available to the bacteria, or more utilization by the faecal flora of the ammonia produced by the deaminating organisms.

Mason, Just & Bech-Andersen (1976) examined the effect of varying the degree of bacterial activity in the large intestines of pigs on the apparent digestibility of amino acids. This was done by replacing maize starch, which is almost wholly digested before the caecum, by potato starch which passes into the large intestines in significant quantities, and also by including or omitting antibiotics. They found that reducing the bacterial fermentation, either by choice of starch or use of antibiotics, significantly increased the digestibility of the α-amino nitrogen and of most of the individual amino acids.

QUANTITATIVE ASPECTS OF PROTEIN DIGESTION

The physiological factors affecting protein digestibility in the pig and its measurement have been reviewed by Rérat, Corring & Laplace (1976). The digestibility of the protein in a diet will depend on the degree to which it is hydrolysed to small peptides and amino acids in the stomach and the lumen of the proximal small intestine, the further hydrolysis and absorption by the mucosa of the small intestine and the digestion and absorption, if any, in the large intestines.

Zebrowska (1973b), in studies on seven 60 kg pigs each fitted with re-entrant cannulas in the proximal duodenum and the distal ileum, reported that throughout the small intestine lumen, the concentration of free amino acids greatly exceeded that of peptides and proteins. In the duodenum, the concentration of peptides exceeded that of proteins, but the reverse was the case in the lower ileum where an appreciable part of the nitrogenous matter consisted of endogenous proteins of low digestibility. Fig. 7.1 shows the relative proportions of the different nitrogenous constituents found by Zebrowska in the duodenum and distal ileum of pigs on different diets.

Pekas, Hays & Thompson (1964) showed that diversion of pancreatic juice by ligating the pancreatic ducts of piglets halved the apparent protein digestibility in 4 week old piglets on a diet with a poorly digestible protein (soybean meal). On piglets of 8 weeks of age, there was only a 20% reduction in digestibility, and, on a milk protein diet, the reduction in digestibility was only 16% at 4 weeks and 13% at 8 weeks. In pancreatic duct diversions on older pigs, Anderson & Ash (1971) found reductions in protein digestibility of 15–17%. Imondi, Stradley & Wohlgemuth (1972), in 12 week old pigs, found that pancreatic ligation reduced protein digestibility to less than half.

Rérat & Corring diverted the pancreatic juice from the intestines in pigs and replaced it by a corresponding salt solution (Rérat, Corring & Laplace, 1976). This halved the amino acid absorption during 10 hours from a fishmeal diet, while diverting the pancreatic juice without the replacement by infusion of the appropriate salt solution, halved it again (Fig. 7.2). Rérat, Corring, Vaissade & Vaugelade (1977) found that the effect of pancreatic diversion on protein digestion, whether measured by protein digestibility or by portal blood amino acid levels, depended on dietary protein levels, being greater on a high protein diet.

While information on the factors influencing absorption and hydrolysis by the mucosa is very scanty, some of the factors affecting peptic and tryptic hydrolysis are well known. As has been described in Chapter 5, the capacity of the piglet to secrete pepsin and acid is limited, and is extremely low at birth. Thus, the gastric digestion of proteins is very much dependent on age. While the capacity of secrete trypsin and chymotrypsin is quite substantial at birth, it nevertheless increases with age, at least in the first few weeks of life.

The action of the pancreatic proteases is also affected by inhibitors which occur in some foods. A heat-labile trypsin inhibitor was reported in raw soyabean meal by Ham & Sandstedt (1944), isolated by Kunitz (1946) and its structure determined by Koide & Ikenaka (1973). It has now been shown that soybeans contain several proteinase inhibitors (Hwang, Foard & Wei, 1977), some resembling the Kunitz trypsin inhibitor (Koide & Ikenaka, 1973) and others of a different type which are more general proteinase inhibitors (Odani & Ikenaka, 1973).

Similar inhibitors have been found in a number of other plant materials sometimes used in pig feeding, mainly in leguminosae, such as the lima bean (Tauber, Kershaw & Wright, 1949) and the Indian field bean and the double bean (Sohonie

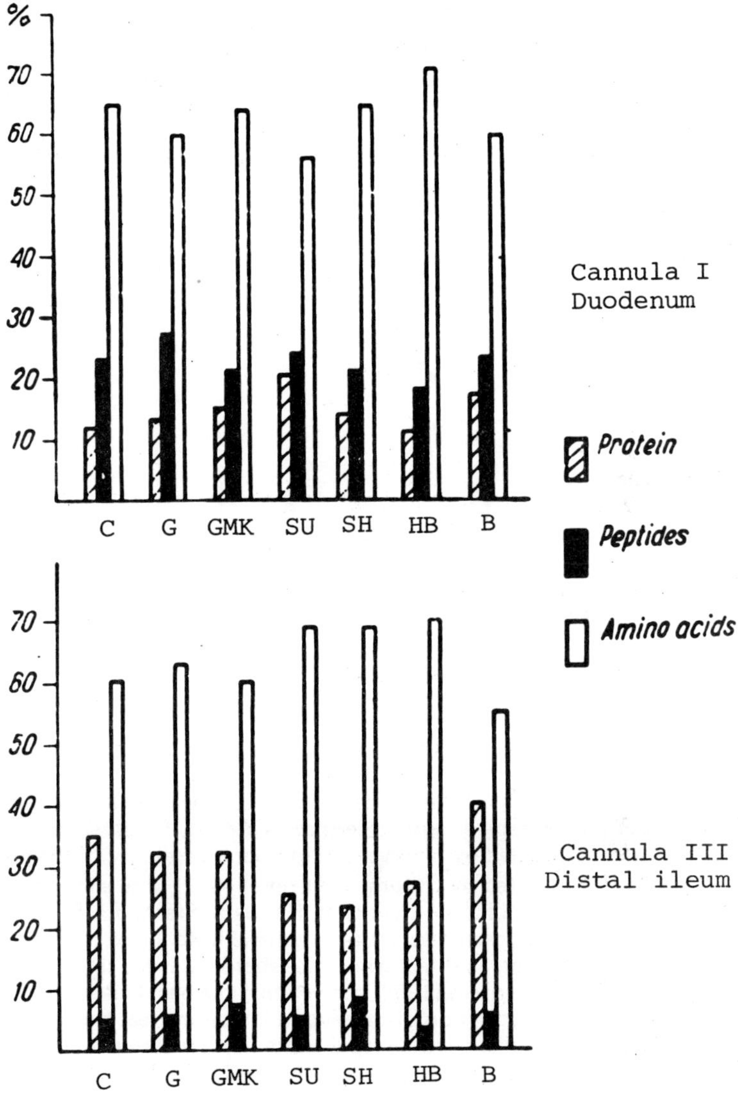

Fig. 7.1 Proportion of protein, peptides and free amino acids in the soluble nitrogen fraction of digesta passing through the duodenum and the distal ileum of pigs fed with the diets: casein (C), gluten (G), control (GMK), soya bean meal, unheated (SU), soya bean meal, heated (SH), horse bean meal (HB) and barley (B). (Zebrowska, 1973b).

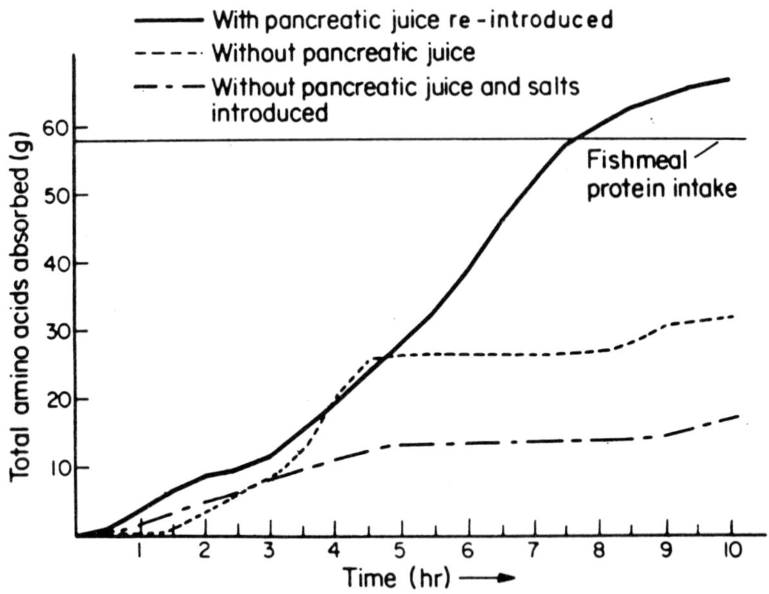

Fig. 7.2 *Absorption of amino acids (g) during 10 hours after the ingestion of 500 g of a semisynthetic meal (containing 58 g fish meal protein). Pig fitted with a permanent cannula of the pancreatic duct and of the duodenum: with or without re-introduction of the pancreatic juice. (Rérat, Corring & Laplace, 1976).*

& Ambe, 1955a). A protease inhibitor has also been found in potato (Sohonie & Ambe, 1955b) which is primarily a chymotrypsin inhibitor (Balls & Ryan, 1963). Like the soybean inhibitor, most of these inhibitors are heat-labile, but the one in the lima bean is heat stable (Tauber, Kershaw & Wright, 1949). Discovery of these inhibitors was initiated by observations of poor digestibility of the protein in raw soybean first reported by Osborne & Mendel (1917), who showed that this was improved by suitable heat-treatment. Hayward, Steenbock & Bohstedt (1935) found that raw soybean meal caused a general lowering of the digestibility of protein in the diet, an effect which was abolished if the soybean was suitably heated, and Muelenaere (1964) showed that the beneficial effect of heating was due to the destruction of the trypsin inhibitor. Heat treatment of soybean meal used in pig diets to improve digestibility has been practised for many years, and, as referred to later, Heinz & Poppe (1975) have studied the effect of various heat treatments on true digestibility in pigs of the protein and individual amino acids in soybean meal. Whittemore, Taylor, Moffat & Scott (1975) have shown that raw, but not heated potatoes lower the digestibility of protein in diets to which they are added.

Sow colostrum contains a powerful trypsin inhibitor which is present at high levels in the first day of lactation (Laskowski, Kassell & Hagerty, 1957). The level of this in the milk falls rapidly and is almost zero by the fifth day. It presumably protects the immune globulins which are absorbed intact during the first day of life of the piglets.

Endogenous Nitrogen

During the passage of food through the alimentary tract, considerable quantities of endogenous protein are added in the form of digestive secretions and desquamated mucosal cells. Very varied estimates have been made of the size of this addition of endogenous protein. Nasset (1965) suggested that it was several times greater than the protein content of the diet, but much lower estimates are usual today.

The endogenous protein can be calculated, either by adding together the quantities, if known, secreted by the various parts of the tract, or by comparing the nitrogen levels in the contents of the tract of pigs on diets with different protein levels.

In listing the information on secretion of the different parts of the tract, Low (1976) pointed out that much of it is very unsatisfactory, having been determined on a few animals of one particular age, often by methods with serious limitations. In particular, many of the secretions vary considerably in their protein content, so that only a study over a considerable period can give results which are useful for calculation of endogenous secretion.

Looking at the individual components of the endogenous protein, Corring (1975) made measurements of pancreatic secretion in 7 pigs of approximately 45 kg. He found the daily output of protein in the pancreatic juice to be 18.6 ± 0.6 g per day, which represented 11% of the dietary nitrogen and 5 to 8% of the amino acid requirement of the pig. Measuring the secretion of intestinal juice into an isolated jejunal loop, on 70 kg pigs, Horszczaruk, Buraczewska & Buraczewski (1974) calculated by extrapolation that the secretion into small intestine from the intestine wall would have been 62–73 g crude protein on the normal protein diet and 50–62 g on the protein-free diet. It is impossible to say whether such a loop was representative, and whether the degree of shedding of cells or other protein into such a loop was abnormal.

On the basis of studies on pigs with duodenal fistulas, fed a diet containing a constant amount of an unabsorbable marker (Cr_2O_3), Horszczaruk (1971a) estimated the endogenous protein to be equal to 20% of the dietary protein on a high protein diet and 30% on a low protein diet. He commented on the considerable variability of composition of the duodenal contents, a fact which may account for the variability of estimates of endogenous protein.

Dammers (1964) estimated the endogenous nitrogen in pigs by feeding a protein-free ration and measuring the faecal nitrogen. A development of this technique is the "regression" technique which involves feeding of rations of different protein levels from as near zero as possible to relatively high levels, and to measure the faecal nitrogen on each diet. Endogenous nitrogen up to different points along the gut has also been measured by this technique. Zebrowska & Buraczewska (1972a), feeding diets with different levels of protein to 50 kg pigs with re-entrant fistulas, found that up to a point just distal to the duodenum, the endogenous nitrogen was 6–7 g per day, whether the pig was on a diet with a high, medium, low or zero protein level. Zebrowska, Buraczewska, Burzczewski & Horszczaruk (1975) gave a similar figure (7 g per day) for the endogenous nitrogen just distal to the duodenum, and a figure of

3 g per day in the distal ileum. A comparable quantity, 5.6 –7.6 g/24 hours endogenous nitrogen, just distal to the duodenum, was obtained by Cuperlovic, Hristić & Zebrowska (1975) in 30 kg pigs on the basis of the difference between dietary intake and the quantity passing through a re-entrant fistula. However, Low & Zebrowska (1977) have shown that this figure for apparent endogenous nitrogen up to the duodenal fistula is materially influenced by the technique of collection, and becomes zero if the collection is made at luminal pressure.

The amino acid composition of the endogenous nitrogen was examined by Zebrowska, Simon, Münchmeyer & Bergner (1976) who compared the composition of collections from duodenal fistulas in 50 kg pigs with that of the diet. They found the endogenous nitrogen to consist predominantly of non-essential amino acids, and especially glycine.

The endogenous protein is itself partly digested and absorbed, so that by the mid small intestine, the ratio of protein to an indigestible marker added to the diet is already less than the ratio in the diet, in spite of the endogenous protein addition (Horszczaruk 1971a). However, the endogenous nitrogen is less digestible than the dietary, as was shown by Zebrowska & Buraczewska (1972a, b), Zebrowska (1973b) and Buraczewska, Buraczewski & Zebrowska (1975). All of these reports indicated that the amino acid composition of the duodenal contents was similar to that of the diet for a range of diets of very different amino acid composition, but that by the terminal ileum, the contents on all the diets had a similar amino acid composition to that of a pig on a non-protein diet. Thus the endogenous nitrogen made a fairly small contribution to the duodenal contents but, being less digestible, provided a large proportion of the nitrogen still unabsorbed by the terminal ileum. The relatively lower digestibility of the endogenous nitrogen was also shown by Zebrowska, Simon, Münchmeyer & Bergner (1976), who injected isotopically labelled leucine and lysine intravenously into 50 kg pigs and took samples at intervals from duodenal and ileal fistulas and from the faeces. There was an increase in the proportion of labelled leucine and lysine along the small intestine.

Using the regression technique for measuring endogenous nitrogen reaching the faeces, Schurig & Poppe (1976a) found that very large differences in the amino acid composition of the diet made very little difference to either the quantity or the composition of the endogenous nitrogen reaching the faeces. They found (Schurig & Poppe, 1976b) that the endogenous nitrogen, as a proportion of food intake, was influenced by body weight, being greater in pigs of 10–40 kg body weight than in pigs of over 40 kg.

Net Digestion in Various Parts of the Gut

Net digestion in various sections of the gut has been measured by Horszcaruk (1971a, b, c), Holmes, Bayley & Horney (1973) and Keys & DeBarthe (1974). Some of the results obtained by Horszczaruk (1971b) are given in Table 7.3. Horszczaruk (1971c) showed that the net protein absorption in the large occurred almost entirely in the proximal half of the colon. Holmes et al. (1973), in agreement with Horszczaruk, showed that the greater part of the overall protein absorption occurred in the small intestine. In diets with an apparent digestibility of nitrogenous matter of 81–87%, they found that the net protein absorption up to the end of the small intestine was 70–78%. Keys & DeBarthe

Table 7.3 Apparent digestibility coefficients up to various points along the gastro-intestinal tract. Mean of 3 pigs. (from Horszczaruk, 1971b).

	Dry matter	Organic matter	Crude protein	True protein
Stomach	−18.8	−19.2	−12.6	13.1
Small intestine I	25.9	29.8	−103.0	−26.8
Small intestine II	62.0	65.1	36.7	55.6
Caecum	71.3	75.3	67.1	68.0
Colon I	73.2	77.1	72.2	68.4
Colon II	77.6	80.7	72.7	73.1
Rectum	77.6	80.7	73.2	73.8

(1974) obtained slightly lower protein digestibilities on milo diets but similar results to the other workers on wheat, corn and barley diets. They found significant differences in the protein digestibilities between the diets in the large intestine but not in the other sections of the gut.

Zebrowska (1973a) and Holmes, Bayley, Leadbeater & Horney (1974) have studied the apparent digestion of individual amino acids in different sections of the gut. Zebrowska (1973c) and Buraczewska, Buraczewski & Zebrowska (1975) showed that the percentage absorption in the small intestine of an amino acid present in dietary protein varied both with the amino acid concerned, and with the diet. Threonine, and, in the earlier experiment, also serine and alanine had the lowest absorptions on all the diets, but which amino acids were best absorbed varied from one diet to another. Holmes et al. (1974) found that net digestion of individual amino acids up to the terminal ileum varied with diet, but that these differences had disappeared by the faeces, being presumably obscured by the bacterial action in the large intestine. In diets containing soya bean meal or rape seed meal as the protein source, they also found relatively low net digestion in the small intestine for threonine, serine, and alanine but found the lowest values for proline and glycine.

Protein and Amino Acid Digestibility

Because of the difficulty of obtaining reliable figures for endogenous nitrogen, most of the overall digestibility measurements for proteins and amino acids have been of apparent digestibility, based on the net percentage decrease in the substance in question on passage of food through the gut (see Chapter 9).

Dammers (1964) measured the apparent digestibility in pigs of total protein and the individual amino acids for a large number of different dietary constituents. He then assessed the endogenous nitrogen by feeding a non-protein diet and measuring the faecal nitrogen and, from this, calculated true digestibilities.

Meier, Poppe & Jentsch (1975) applied the regression technique for measuring protein digestibility in pigs using a series of diets differing in protein content from almost zero to quite high levels, and calculating the true digestibility of nitrogen, protein and individual amino acids. They found the results to be very similar to those obtained using classical methods (Meier, Jentsch, Jancker & Poppe, 1974). This regression technique was applied by Heinz & Poppe (1975) to measure the true digestibility in 10–18 kg pigs of the nitrogen, protein and amino acids on diets containing soybean meal which had been subjected to a variety of heat treatments. In the diets containing unheated soya meal their value for true nitrogen digestibility was 77.8%, while digestibility of the essential amino acids ranged from 71% for methionine to 89.8% for arginine. For the optimally heated meal the figures were 92.3% for nitrogen, 89.5% for methionine and 97% for arginine. Similar studies with horse-beans by Meier, Heinz & Poppe (1975) give a similar pattern of relative digestibilities for the different amino acids, but a high digestibility in the raw product which was reduced by heat treatment.

Schurig & Poppe (1976b) using the same technique with barley rations, obtained a comparable pattern of true amino acid digestibilities ranging from about 82% for arginine to about 70% for methionine. When they made corrections for the different levels of endogenous nitrogen in pigs of different live weight, they found that the true digestibilities of nitrogen and amino acids appeared to be independent of live weight over the range 10–100 kg.

Zebrowska, Simon, Münchmeyer & Bergner (1976) showed the difference between the digestibilities of different amino acids as measured in the faeces and in the distal ileum and, in view of the loss of amino acids by bacterial action in the large intestines, questioned the validity of digestibility measurements of amino acids measured at the faecal level as an indication of their usefulness to the pig.

REFERENCES

Abita, J.P., Delaage, M., Lazdunski, M. & Savdra, J. (1969) The mechanism of activation of trypsinogen. The role of the four N-terminal aspartyl residues. *Eur. J. Biochem., 8,* 314–324.

Al-Janabi, J., Hartsuck, J.A. & Tang, J. (1972) Kinetics and mechanism of pepsinogen activation. *J. biol. Chem., 247,* 4628–4632.

Anderson, D.M. & Ash, R.W. (1971) The effect of ligating the pancreatic duct on digestion in the pig. *Proc. Nutr. Soc., 30,* 34A–35A

Ardelt, W. (1974) Partial purification and properties of porcine pancreatic elastase II. *Biochim. biophys. Acta, 341,* 316–326.

Ardelt, W. (1975) Physical parameters and chemical composition of porcine pancreatic elastase II. *Biochim. biophys Acta, 393,* 267–273.

Balls, A.K. & Ryan, C.A. (1963) Concerning a crystalline chymotryptic inhibitor from potatoes, and its binding capacity for the enzyme. *J. biol. Chem., 238,* 2976–2982.

Baló, J. & Banga, I. (1949) Die Zerstörung der elastischen Fasern der Gefässwand. *Schweiz. Z. allg. Path. Bact., 12,* 350–361.

Baló, J. & Banga, I. (1950). The elastolytic activity of pancreatic extracts. *Biochem. J., 46,* 384–387.

Barns, R.J. & Elmslie, R.G. (1977) Identification of a mucosal form of enteropeptidase in triton X-100 extracts of porcine duodenal mucosa. *Biochim biophys Acta., 480,* 450–460.

Beyer, M.G., Steinhart, H. & Kirchgessner, M. (1975) Zur Auswirkung verschiedener Enzym-Substrat-Verhältnisse auf Molekulargewicht und Aminosäurenzusammensetzung der bei der peptischen in-vitro Verdauung von Proteinen entstehenden Peptide. *Landwirtsch. Forsch., 28,* 269–277.

Bovey, F.A. & Yanari, S.S. (1960) Pepsin *in* The enzymes. Edn 2, vol. 4. Eds Boyer, P.D., Lardy, H. & Myrbäck, K., pp. 63–91. Academic Press, New York.

Braude, R., Cutts, I.R., Myres, A.W. & Porter, J.W.G. (1970) Measurement of amino acid levels in portal blood of pigs and of rate of portal blood flow. *Proc. Nutr. Soc., 29,* 60A.
Braude, R., Mitchell, K.G., Myres, A.W., Porter, J.W.G. & Williams, A.P. (1969) Amino acid levels in blood plasma of growing pigs given diets supplemented with lysine. *Proc. Nutr. Soc., 28,* 40A–41A.
Buraczewska, L., Buraczewski, S. & Zebrowska, T. (1975) Digestion and absorption in the small intestine of Pigs. 2. Amino acid content in digesta and their absorption. *Roczn. Nauk Roln., 97-B-1,* 103–115.
Buraczewska, L. & Zebrowska, T. (1972) Wchlanianie aminokwasów z kaseiny i jej hydrolizatu w izolowanym odcinku jelita cienkego świń. *Zesz. Probl. postep. Nauk Roln., 126,* 129–133.
Carnegie, P.R. (1971) Amino acid sequence of the encephalitogenic basic protein from human myelin. *Biochem. J., 123,* 57–67.
Chachulowa, J. (1964) Wplyw niektórych dodatków do paszy na poziom ammoniaku w jelicie świń.
1. Zawartość ammoniaku w jelicie ślepym świń przy róznym dodatku celulozy do paszy, chlortetracykliny i siarczanu miedzi.
Roczn. Nauk Roln., 84-B-1, 83–95.
Charles, M., Rovery, M., Guidoni, A. & Desnuelle, P. (1963) Sur le trypsinogène et la trypsine de porc. *Biochim. biophys. Acta, 69,* 115–129.
Corring, T. (1975) Apport de protéines d'origine endogène par la sécrétion du pancréas exocrine chez le porc. *Annls Biol. anim. Biochim. Biophys., 15,* 115–118.
Corring, T. & Saucier, R. (1972) Sécrétion pancréatique sur porc fistulé. Adaptation a la teneur en protéines du régime. *Annls Biol. anim. Biochim. Biophys., 12,* 233–241.
Craft, I.L., Geddes, D. & Matthews, D.M. (1968) Observations on the intestinal absorption of glycine peptides in man. *J. Physiol., Lond., 196,* 31P–32P.
Cuperlović, M., Hristić, V. & Zebrowska, T. (1975) Influence of the composition of the dietary protein supplement on the digestion and absorption of protein of the small intestine of pigs. *Acta Veterinaria, Belgrade, 25,* 287–300.
Dammers, J. (1964) Verteringsstudies bij het varken. Thesis, Leuven.
Delort-Laval, J., Charlet-Lery, G. & Zelter, S.Z. (1963) Efficacité de quelques protides alimentaires chez le porc. *Annls Biol. anim. Biochim. Biophys., 3,* 369–380.
Desnuelle, P., Rovery, M. & Bonjour, G. (1950) Quelques nouvelles observations sur le mode d'attaque de la globine de cheval et de l'albumine d'oeuf par la pepsine. *Biochim. biophys. Acta, 5,* 116–137.
Etherington, D.J. & Taylor, W.H. (1971) The separation and preparation of human pepsins 3 and 5 from the gastric juice of single individuals and the mode of action of human and swine pepsins on B chain of oxidised insulin. *Biochim. biophys. Acta, 236,* 92–98.
Fauconneau, G. & Michel, M.C. (1970) The role of the gastrointestinal tract in the regulation of protein metabolism. *in* Mammalian Protein Metabolism, Vol. 4 Ed. Munro, H.N. pp 481–522, Academic Press, New York.
Fisher, R.S. (1967) Absorption of proteins. *Proc. Nutr. Soc., 26,* 23–27.
Folk, J. E. & Cole, P. W. (1965) Chymotrypsin C. II Enzymic specificity towards several polypeptides. *J. biol. Chem., 240,* 193–197.
Folk, J.E. & Schirmer, E.W. (1963) The porcine pancreatic carboxypeptidase A system. I. Three forms of the active enzyme. *J. biol. Chem., 238,* 3884–3894.
Fritz, H., Hüller, I., Wiedemann, M. & Werle, E. (1967) Über Proteaseinhibitoren. V. Zur Chemie und Physiologie der spezifischen Trypsininhibitoren aus der Bauchspeicheldrüsen von Rind, Hund, Schwein and Mensch. *Hoppe-Seyler's Z. physiol. Chem., 348,* 405–418.
Hadorn, B., Steiner, N., Sumida, C. & Peters, T.J. (1971) Intestinal enterokinase. Mechanisms of its "secretion" into the lumen of the small intestine. *Lancet, i,* 165–166.
Ham, W.E. & Sandstedt, R.M. (1944) A proteolytic inhibiting substance in the extract from unheated soy bean meal *J. biol. Chem., 154,* 505–506.
Harmeyer, J., Martens, H. & Birck, R. (1974) Eine in-vitro-Methode zur Messung intestinaler Absorptionsvorgänge bei Ferkeln-Messung der Aminosäurenaufname. *Wien. Tierärztl. Mschr., 61,* 178–184.
Haylett, T., Swart, L.S. & Parris, D. (1971) Studies on the high-sulphur proteins of reduced merino wool. Amino acid sequence of protein SCMKB-III. B3. *Biochem J., 123,* 191–200.
Hayward, J.W., Steenbock, H. & Bohstedt, G. (1935) The effect of heat as used in the extraction of soybean oil upon the nutritive value of the protein of soybean oil meal. *J. Nutr., 11,* 219–234.

Heinz, T. & Poppe, S. (1975) Der Einfluss thermischer und mechanischer Behandlung von Futtermitteln duf die wahre Verdaulichkeit der Aminosäuren.
1. Die wahre Verdaulichkeit der Aminosäuren bei unterschiedlich hitzbehandelten Sojaextraktionsschroten. *Arch. Tierernähr., 25,* 353–358.
Holmes, J.H.G., Bayley, H.S. & Horney, F.D. (1973) Digestion and absorption of dry and high moisture diets in the small and large intestine of the pig. *Br. J. Nutr., 30,* 401–410.
Holmes, J.H.G., Bayley, H.S., Leadbeater, P.A. & Horney, F.D. (1974) Digestion of protein in small and large intestine of the pig. *Br. J. Nutr., 32,* 479–489.
Holmes, R. & Lobley, R.W. (1970) The localization of enterokinase to the brush-border membrane of guinea-pig small intestine. *J. Physiol., Lond., 211,* 50–51.
Horszczaruk, F. (1971a) Badania nad trawieniem u świń. III. Badania nad trawieniem i wchlanianiem prostych i mostkowych przetok jelita cienkego. *Biul. Inst. Genetiki Hodowli Zwierzat PAN, 21,* 137–172.
Horszczaruk, F. (1971b) Badania nad trawieniem u świń. I. Badania z zastosowaniem przetok jelita ślepego i okreznici. *Biul. Inst. Genetiki Hodowli Zwierzat PAN, 21,* 101–115.
Horszczaruk, F. (1971c) Badania nad trawieniem u świń. II. Badania nad trawieniem i wchlanianiem przy zastosowaniu metody ubojowej. *Biul. Inst. Genetiki Hodowli Zwierzat PAN, 21,* 117–136.
Horszczaruk, F., Buraczewska, L. & Buraczewski, S. (1974) Ilość i sklad soku jelitowego wydzielanego do isolowanej petli jelita cienkiego u świń. *Roczn. Nauk Roln., 95-B-4,* 69–77.
Hwang, D.L., Foard, D.E. & Wei, C.H. (1977) A soybean trypsin inhibitor. Crystallization and X-ray crystallographic study. *J. biol. Chem., 252,* 1099–1101.
Imondi, A.R., Stradley, R.P. & Wohlgemuth, R. (1972) Enzyme replacement therapy in the pancreatic duct ligated swine. *Proc. Soc. exp. Biol., 141,* 367–372.
Inouye, K. & Fruton, J.S. (1967) Studies on the specificity of pepsin. *Biochemistry, 6,* 1765–1777.
Irie, M. (1968) A comparative study of several pancreatic ribo-nucleases: cleavage of dinucleoside phosphates and nucleoside cyclic phosphates. *J. Biochem. Tokyo, 63,* 501–505.
Jones, W. (1912) On the formation of guanylic acid from yeast nucleic acid. *J. biol. Chem., 12,* 31–35.
Jones, W. (1920) The action of boiled pancreas extract on yeast nucleic acid. *Am. J. Physiol., 52,* 203–207.
Josefsson, L. & Lindberg, T. (1965a) Intestinal dipeptidases. I. Spectrophotometric determination and characterization of dipeptidase activity in pig mucosa. *Biochim. biophys. Acta, 105,* 149–161.
Josefsson, L. & Lindberg, T. (1965b) Intestinal dipeptidases. II. Distribution of dipeptidases in the small intestine of the pig. *Biochim. biophys. Acta, 105,* 162–166.
Keys, J.E. & DeBarthe, J.V. (1974) Site and extent of carbohydrate, dry matter, energy and protein digestion and the rate of passage of grain diets in swine. *J. Anim. Sci., 39,* 57–67.
Kim, Y.S., Kim, Y.W. & Sleisenger, M.H. (1974) Studies on the properties of peptide hydrolases in the brush border and soluble fractions of small intestinal mucosa of rat and man. *Biochim. biophys. Acta., 370,* 283–296.
Kirchgessner, M., Beyer, M.G. & Steinhart, H. (1976) Activation of pepsin (EC3.4.4.1) by heavy metal ions including a contribution to the mode of action of copper sulphate in pig nutrition. *Br. J. Nutr., 36,* 15–22.
Koide, T. & Ikenaka, T. (1973) Studies on soybean trypsin inhibitors.
3. Complete amino acid sequence of the carboxyl-terminal region and the complete amino acid sequence of soybean trypsin inhibitor (Kunitz). *Eur. J. Biochem, 32,* 417–431.
Kratzer, F.H. & Porter, J.W.G. (1962) The effect of pH on the digestion of proteins in vitro by pepsin. *Br. J. Nutr., 16,* 579–584.
Kunitz, M. (1946) Crystalline soybean trypsin inhibitor. *J. gen. Physiol., 29,* 149–154.
Kunitz, M. & Northrop, J.H. (1936) Isolation from beef pancreas of crystalline trypsinogen, trypsin, a trypsin inhibitor and an inhibitor-trypsin compound. *J. gen. Physiol., 19,* 991–1007.
Laskowski, M., Kassell, B. & Hagerty, G. (1957) A crystalline trypsin inhibitor from swine colostrum. *Biochim. biophys. Acta, 24,* 300–305.
Lawrence, T.L.J. (1972) The effect of certain dietary factors on in vivo pH changes and pepsin activity in the stomach of the growing pig. *Br. vet. J., 128,* 402–411.
Ledoux, M. & Lamy, F. (1975) Electrophoretic characterization of porcine pancreatic (pro)elastase A and B. *Can. J. Biochem., 53,* 421–432.

Lee, D. & Ryle, A.P. (1967a) Pepsinogen D. A fourth proteolytic zymogen from pig gastric mucosa. *Biochem. J., 104,* 735–741.
Lee, D. & Ryle, A.P. (1967b) Pepsin D. A minor component of commercial pepsin preparations. *Biochem. J., 104,* 742–748.
Louvard, M.N. & Puigserver, A. (1974) On bovine and porcine anionic trypsinogens. *Biochim. biophys. Acta, 371,* 177–185.
Low, A.G. (1976) Digestion and absorption of nutrients in growing pigs. *Proc. Nutr. Soc., 35,* 57–67.
Low, A.G. & Zebrowska, T. (1977) Dry matter and nitrogen in the duodenal contents of growing pigs: a discrepancy explained. *Br. J. Nutr., 38,* 145–147.
Marchis-Mouren, G. (1959) Contribution a l'étude de la purification de la lipase de porc et a l'étude de l'équipement enzymatique du suc pancréatique de diverses espèces. Thesis, Marseille.
Marchis-Mouren, G. (1965) Etude comparée de l'équipement enzymatique du suc pancréatique de diverses espèces. *Bull. Soc. Chim. biol. 47,* 2207–2217.
Marchis-Mouren, G., Charles, M., Ben Abdeljlil, A. & Desnuelle, P. (1961) Sur l'équipement en enzymes du suc pancréatique de porc et de chien. *Biochim. biophys. Acta, 50,* 186–188.
Maroux, S., Baratti, J. & Desnuelle, P. (1971) Purification of porcine enterokinase. *J. biol. Chem., 246,* 5031–5039.
Maroux, S., Louvard, D. & Baratti, J. (1973) The aminopeptidase from hog intestinal brush border. *Biochim. biophys. Acta, 321,* 282–295.
Maroux, S. & Louvard, D. (1977) Les hydrolases de la bordure en brosse des entérocytes de porc. Leur mode d'intégration a travers l'étude de l'aminopeptidase. *Gastroenterol. Clin. Biol., 1,* 377–388.
Mason, V.C., Just, A. & Bech-Andersen, S. (1976) Bacterial activity in the hindgut of pigs. 2. Its influence on the apparent digestibility of nitrogen and amino acids. *Z. Tierphysiol. Tierernähr. Futtermittelk., 36,* 310–324.
Matthews, D.M. (1971) Protein absorption. *J. clin. Path., 24, suppl. (Roy. Coll. Path.), 5,* 29–40.
Matthews, D.M. (1975) Intestinal absorption of peptides. *Physiol. Rev., 55,* 537–608.
Matthews, D.M., Craft, I.L., Geddes, D.M., Wise, I.J. & Hyde, C.W. (1968) Absorption of glycine and glycine peptides from the small intestine of the rat. *Clin. Sci., 35,* 415–424.
Matthews, D.M., Lis, M.T., Cheng, B. & Crampton, R.F. (1969) Observations on the intestinal absorption of some oligopeptides of methionine and glycine in the rat. *Clin. Sci., 37,* 751–764.
Meier, H., Heinz, T. & Poppe, S. (1975) Der Einfluss thermischer und mechanischer Behandlung von Futtermitteln auf die wahre Verdaulichkeit der Aminosäuren. 2. Die wahre Verdaulichkeit der Aminosäuren von behandelten Ackerbohnen. *Arch. Tierernähr., 25,* 359–364.
Meier, H., Jentsch, H., Jancker, H. & Poppe, S. (1974) Untersuchungen zur Additivität von Aminosäuren-resorbierbarkeitswerten an Schweinen. 1. Untersuchungen nach dem klassischen Verfahren. *Arch. Tierernähr., 24,* 385–396.
Meier, H., Poppe, S. & Jentsch, H. (1975) Untersuchungen zur Additivät von Aminosäuren-resorbierbarkeitswerten an Schweinen. 2. Untersuchungen nach dem regressiven Verfahren. *Arch. Tierernähr., 25,* 47–58.
Messerli, H. (1913) Uber die Resorptionsgeschwindigkeit der Eiweiss und ihrer Abbauprodukte im Dünndarm. *Biochem. Z., 54,* 446–473.
Michel, M.C. (1966) Métabolisme de la flore intestinale du porc. Dégradation de formes L et D des acides aminés. *Private coll. Soc. org. biol. Chem., Paris, Cahier 6,* 307–324.
Michel, M.C., Jouandet, C., Salmon-Legagneur, E., Aumaitre, A. & Francois, A.C. (1964) Influence de l'acrylate de sodium sur la croissance du porcelet. *Annls Zootech., 13,* 341–350.
Moring-Claesson, I. (1948) Frontal analysis studies on the adsorption of crystalline egg albumen and its cleavage products. *Biochim. biophys. Acta, 2,* 386–408.
Muelenaere, H.J.H. (1964) Studies on the digestion of soybeans. *J. Nutr., 82,* 197–205.
Narayanan, A.S. & Anwar, R.A. (1969) The specificity of purified porcine elastase. *Biochem. J., 114,* 11–17.
Nasset, E.S. (1965) Role of the digestive system in protein metabolism. *Fedn Proc. Fedn Am. Socs exp. Biol., 24,* 953–958.

Nehring, K. & Schröder, I. (1965) Untersuchung zur enzymatischen Hydrolyse von Eiweissfutterstoffen in vitro.
4. Über die enzymatische Freisetzung der Aminosauren aus Eiweissfutterstoffen. *Arch. Tierernähr., 15,* 113–134.

Nordström, C. & Dahlqvist, A. (1970) The cellular localization of enterokinase. *Biochim. biophys. Acta, 198,* 621–622.

Nordström, C. & Dahlqvist, A. (1971) Rat enterokinase: The effect of ions and the localization in the intestine. *Biochim. biophys. Acta, 242,* 209–225.

Norén, O., Dabelsteen, E., Sjöström, H. & Josefsson, L. (1977) Histological localization of two peptidases in the pig small intestine and liver, using immunofluorescence. *Gastroenterology, 72,* 87–92.

Norén, O., Sjöström, H. & Josefsson, I. (1973) Studies on a soluble dipeptidase from pig intestinal mucosa. I. Purification and specificity. *Biochim. biophys. Acta, 327,* 446–456.

Odani, S. & Ikenaka, T. (1973) Studies on soybean trypsin inhibitors. VIII Disulphide bridges in soybean Bowman-Birk proteinase inhibitor. *J. Boichem., Tokyo, 74,* 697–715.

Orok, E.J. & Bowland, J.P. (1975) Rapeseed, peanut and soybean meals as protein supplements: dietary amino acid composition and postprandial plasma amino acid pattern as indices of performance in pigs. *Can. J. Anim. Sci., 55,* 337–346.

Osborne, T.B. & Mendel, L.B. (1917) The use of soy bean as food. *J. biol. Chem., 32,* 369–376.

Pekas, J. C., Hays, V.W. & Thompson, A.M. (1964) Exclusion of the exocrine pancreatic secretion: effect on digestibility of soybean and milk protein by baby pigs at various ages. *J. Nutr., 82,* 277–286.

Peters, T.J. (1970) The subcellular localization of di- and tri-peptide hydrolase activity in guinea-pig small intestine. *Biochem. J. 120,* 195–203.

Pion, R., Fauconneau, G. & Rérat, A. (1964) Variation de la composition en acides aminés du sang porte au cours de la digestion chez le porc. *Annls Biol. anim. Biochim. Biophys., 4,* 383–402.

Pion, R., Fauconneau, G. & Rérat, A. (1966) Étude de la digestion des protéines chez le porc par la mesure de l'aminoacidémie porte. *Private Coll. Soc. org. biol. Chem., Paris, Cahier 6,* 328–339.

Polivoda, D.I. & Gritsenko, H.M. (1973) Usvoenie kristallicheskikh aminokislot u protsessy pishchevareniya u swinei. *Sel'khoz. Biol., 8,* 108–112.

Rensmann, F.W. (1976) Zur Bedeutung wechselnder Rohfaser-und Stärkemengen im Futter für den Gehalt Trockensubstanz, flüchtigen Fettsäuren, Milchsäure und Ammoniak im Kot von Ferkeln. Diss, Hannover.

Rérat, A., Corring, T. & Laplace, J.P. (1976) Protein digestion and absorption. *in* Protein metabolism and Nutrition. Eds Cole, D.J.A., Boorman, K.N., Buttery, P.J., Lewis, D., Neale, R.J. & Swan, H. European Association for Animal Production, Publication no. 16. Butterworth, London.

Rérat, A., Corring, T., Vaissade, P. & Vaugelade, P. (1977) Postprandial absorption of α-amino nitrogen and sugars in pigs. Quantitative measurement using discontinuous pancreatic deprivation. *Annls Biol. anim. Biochim. Biophys., 17,* 583–588.

Rérat, A. & Duée, P.H. (1972) Mesure quantitative de l'absorption des acides aminés chez le porc recevant des régimes a taux variable de protéines. *Annls Biol. anim. Biochim. Biophys., 13,* 788–791.

Ryle, A.P. (1960) Parapepsinogen II; The zymogen of parapepsin II. *Biochem. J., 75,* 145–150.

Ryle, A.P. & Porter, R.R. (1959) Parapepsins. Two proteolytic enzymes associated with porcine pepsin. *Biochem. J. 73,* 75–86.

Sanny, C.G., Hartsuck, J.A. & Tang, J. (1975) Conversion of pepsinogen to pepsin. Further evidence for intramolecular and pepsin-catalysed activation. *J. biol. Chem., 250,* 2635–2639.

Schurig, J. & Poppe, S. (1976a) Über die Konstanz der Darmverlustaminosäuren in Abhangigkeit von der Aminosäurenzusammensetzung der Ration. *Arch. Tierernähr., 26,* 475–482.

Schurig, J. & Poppe, (1976b) Zur Bestimmung der wahren Verdaulichkeit von Aminosäuren beim Schwein. *Arch. Tierernähr., 26,* 409–416.

Shimada, M.A. & Zimmerman, D.R. (1973) Hepatic portal catheterization and comparison of portal and anterior vena cava free plasma amino acid patterns in the baby pig. *J. Anim. Sci., 36,* 245–251.
Shotton, D.M. & Hartley, B.S. (1973) Evidence for the amino acid sequence of porcine pancreatic elastase. *Biochem. J., 131,* 643–673.
Sjöström, H. & Norén, O. (1974) Structural properties of pig intestinal proline dipeptidase. *Biochim. biophys. Acta, 359,* 177–185.
Sjöström, H., Norén, O. & Josefsson, I. (1973) Purification and specificity of pig intestinal prolidase. *Biochim. biophys. Acta, 327,* 457–470.
Sohonie, K. & Ambe, K.S. (1955a) Crystalline trypsin inhibitors from the Indian field bean and the double bean. *Nature, Lond., 175,* 508–509.
Sohonie, K. & Ambe, K.S. (1955b) Crystalline inhibitors of trypsin from potato. *Nature, Lond., 176,* 972.
Sokolovsky, M. & Zisapel, N. (1971) Porcine carboxypeptidase B. I. Affinity chromatography and specificity. *Biochim. biophys. Acta, 250,* 203–206.
Steinhart, H. & Kirchgessner, M. (1973) Peptidverteilung auf Molekulargewichts blöcke bei der In-vitro-verdauung von Sojaprotein mit Pepsin unter dem Einfluss verschiedener Verdauungszeiten. *Z. Tierphysiol. Tierernähr. Futtermittelk., 32,* 55–64.
Tauber, H., Kershaw, B.B. & Wright, R.D. (1949) Studies on the growth inhibitor fraction of lima beans and isolation of a crystalline heat-stable trypsin inhibitor. *J. biol. Chem., 179,* 1155–1161.
Taylor, W.H. (1959a) Studies on gastric proteolysis. I. The proteolytic activity of human gastric juice and pig and calf gastric mucosal extracts below pH 5. *Biochem. J., 71,* 73–83.
Taylor, W.H. (1959b) Studies on gastric proteolysis. 4. Proteinase activity of gastric juice and gastric-mucosal extracts at pH 6 to 8. *Biochem. J., 71,* 627–632.
Taylor, W.H. (1962) Proteinases of the stomach in health and disease. *Physiol. Rev., 42,* 519–553.
Tiselius, A. & Eriksson-Quensel, I.B. (1939) Some observations on peptic digestion of egg albumen. *Biochem. J., 33,* 1752–1756.
Travis, J. (1967) Specificity of porcine trypsin. *Biochem. biophys. Res. Commun., 29,* 294–297.
Voytek, P. & Gjessing, E.C. (1971) Studies of an anionic trypsinogen and its active enzyme from porcine pancreas. *J. biol. Chem., 246,* 508–516.
Whittemore, C.T., Taylor, A.G., Moffat, I.W. & Scott, A. (1975) Nutritive value of raw potato for pigs. *J. Sci. Fd Agric., 26,* 255–260.
Yamasaki, Y., Murakami, K., Irie, M. & Ukita, T. (1968) Purification and properties of porcine pancreatic ribonuclease. *J. Biochem., Tokyo, 63,* 25–32.
Zebrowska, T. (1973b) Influence of dietary protein source on the rate of digestion in the small intestine of pigs. I. Amount and composition of digesta. *Roczn. Nauk Roln., 95-B-1,* 115–133.
Zebrowska, T. (1973a) Influence of dietary protein source on the rate of digestion in the small intestine of pigs. II. The rate of protein digestion and amino acid absorption. *Roczn. Nauk Roln., 95-B-1,* 135–155.
Zebrowska, T. (1973c) Digestion and absorption of nitrogenous compounds in the large intestine of pigs. *Roczn. Nauk Roln. 95-B-3,* 85–90
Zebrowska, T. (1975) The apparent digestibility of nitrogen and individual amino acids in the large intestine of pigs. *Roczn, Nauk Roln., 97–B–1,* 117–123.
Zebrowska, T. & Buraczewska, L. (1972a) Wplyw poziomu bialka w diecie na ilość i sklad tresci jelita cienkego świń. *Zesz. Probl. postep. Nauk Roln., 126,* 115–121.
Zebrowska, T., Buraczewska, L., Buraczewski, S. & Horszczaruk, F. (1975). Digestion and absorption in the small intestine of pigs. I. Digestion and absorption of dry matter and nitrogen. *Roczn. Nauk Roln., 96-B-3,* 79–90.
Zebrowska, T. & Buraczewska, L. (1972b) Wplyw poziomu bialka w diecie na przebeig trawienia w jelicie cienkim świń. II. Szybkość trawienia bialka i wchlaniania aminokwasow. *Roczn. Nauk Roln., 94-B-1,* 97–109.
Zebrowska, T., Simon, O., Münchmeyer, R. & Bergner, H. (1976) Untersuchungen zur Sezernierung endogener Aminosäuren in den Verdauungstrakt und zur Aminosäuren-resorption beim Schwein. *Arch. Teirernähr., 26,* 69–82.
Zendzian, E.N. & Barnard, E.A. (1967) Distributions of pancreatic ribonuclease, chymotrypsin and trypsin in vertebrates. *Arch. Biochem., 122,* 699–713.

CHAPTER 8
DIGESTION OF FATS

SUMMARY

Dietary fats consist mainly of triglycerides with some phospholipids, sterols and sterol esters. Endogenous phospholipids are also added to the gut contents, mainly in the bile.

Triglycerides are hydrolysed in the small intestine by the lipase in the pancreatic juice, and the monoglycerides and free fatty acids produced pass into micellar solution in the presence of the bile salts. The monoglycerides and fatty acids are taken up by the mucosa from this micellar solution, the shorter chain acids being transported through the mucosa to the portal blood. The monoglycerides and the longer chain fatty acids are synthesised within the mucosal cells to triglycerides and phospholipids which are accumulated as lipoprotein-covered droplets "chylomicrons" and passed into the intestinal lymph.

DIGESTION IN THE LUMEN

The processes involved in fat digestion in the small intestine lumen have been elucidated by the study of the composition of the digesta and by in vitro studies on the pancreatic lipases and colipases which are secreted into the lumen. The in vitro studies have usually been made using enzymes and coenzymes from pig pancreas. The studies in the lumen have been made usually on man or the rat, but a few carried out on pigs have confirmed that the picture is similar in the pig to that in man and the rat.

The enzymes involved in fat digestion in the lumen of the small intestine are listed in Table 8.1.

Hydrolysis

Triglycerides

The degree of breakdown of triglycerides during digestion, the form in which the products are absorbed and their fate after absorption has only been known since about 1964. Since then, very many review articles have been written on the subject, such as those of Dawson (1971) and Johnston (1968).

Triglycerides are the main components of dietary fats. Their digestion starts in the lumen of the small intestine where they are acted upon by the lipase from the pancreatic juice. This enzyme will only act with any speed where there are separate water and lipid phases present (Sarda & Desnuelle, 1958). The activity of lipase is influenced by the presence of bile salt (Borgström 1954a), and colipase, a co-factor present in pancreatic juice (Erlanson, Fernlund & Borgström, 1973). Borgström (1975) showed that in the absence of colipase or bile salt, the lipase was adsorbed and denatured at the interface and that in the presence of bile salt and absence of colipase, the lipase remained mainly in the aqueous phase so that hydrolysis was inhibited, but that in the presence of both bile salt and colipase, the lipase was adsorbed at the interface, was stable and active. The

Table 8.1 *Pancreatic enzymes involved in fat digestion*

Enzyme	Enzyme Commission name and number	Reactions catalysed	Cofactors for optimum action	Other requirements for optimum action
Lipase	Triacylglycerol lipase (3.1.1.3)	triglyceride → diglyceride ← + fatty acid diglyceride → monoglyceride ← + fatty acid	colipase	separate lipid phase
Phospholipase	Phospholipase A_2 (3.1.1.4)	lecithin → lysolecithin + fatty acid	—	activation by trypsin (secreted as pro-enzyme)
Cholesterol esterase	Cholesterol esterase (3.1.1.13)	cholesterol ester → cholesterol + fatty acid	bile salts	molecular solution

experiments of Borgström & Donnér (1975, 1976), Borgström (1975) and Momsen & Brockman (1976a, b) make it clear that bile salts readily complex with both lipase and colipase, and that when they are all present, a complex of the three is formed which adsorbs at a lipid water interface and is catalytically active.

Lipase acts on esters of primary alcohols (Balls and Matlack, 1938), and so only hydrolyses the ester groups on the 1 and 3 positions of the glycerol moiety, giving first 1, 2-diglycerides and then 2-monoglycerides with the release of free fatty acids (Schonheyder & Volqvartz, 1954; Mattson & Beck, 1956). Both these reactions are readily reversible so that considerable exchange of fatty acids in the 1 and 3 positions occurs during hydrolysis (Borgström, 1954b) and a situation is soon reached where approximately equimolar concentrations of mono-, di- and triglyceride are present (Borgström, 1964). Continued net hydrolysis will then only occur if the hydrolysis products are removed. In the intestine lumen, the monoglycerides and free fatty acids are continuously being removed by the mucosa, so breakdown of tri- and diglyceride continues. Some isomerization of diglyceride and monoglyceride occurs, with migration of acyl groups from the 2 position to the 1 or 3 position, rendering all the ester bonds susceptible to lipase hydrolysis. This occurs more rapidly in glycerides containing short-chain fatty acids, and lipase releases glycerol much more rapidly from these glycerides than from those containing only long-chain fatty acids (Benzonana, Entressangles, Marchis-Mouren, Paséro, Sarda and Desnuelle, 1964). 1-Monoglycerides, provided they are dissolved in the oil phase, are hydrolysed by pancreatic lipase, but in addition, when they are in micellar solution in the presence of bile salts they are hydrolysed by pancreatic cholesterol esterase (Morgan, Barrowman, Filipek-Wender & Borgström, 1968). Thus some complete hydrolysis to glycerol and free fatty acid can occur. Entressangles, Sari & Desnuelle (1966) showed that the rate of apparent hydrolysis of the 2 position of labelled triglyceride by purified pig pancreatic lipase corresponded closely to the rate of isomerization of the partial glycerides. The final stage of hydrolysis to glycerol is not reversible under physiological conditions (Borgström, 1954b). The processes are summed up in Fig. 8.1.

Estimates of the extent of total hydrolysis to glycerol have been made in the rat, where Mattson & Volpenhein (1964) calculated it to be 22% of the ingested triglyceride, while Raghavan & Ganguly (1969) estimated that half an hour after injection into the duodenal lumen, some 14% of the triglyceride had been hydrolysed to glycerol.

Pancreatic lipase will hydrolyse esters of mono, di- and trihydric alcohols being most effective with glycerol (Mattson & Volpenhein, 1972). Apparently conflicting reports have been published on its specificity with regard to the fatty acid component of these esters. The subject has been reviewed in detail by Desnuelle & Savary (1963). The question is confused by the fact, already mentioned, that the physical state of the substrate is important, effective hydrolysis only occurring if it is dispersed as a separate oil phase (Sarda & Desnuelle, 1958). This will only occur with the partially water-soluble esters such as triacetin and dipropionin at high substrate concentrations. At the other end of the scale, esters such as glyceryltristearate which are solid at the reaction temperature, and so are dispersed solids not liquids, are rather slowly hydrolysed, but if stearate is present as one component of a mixed triglyceride of lower

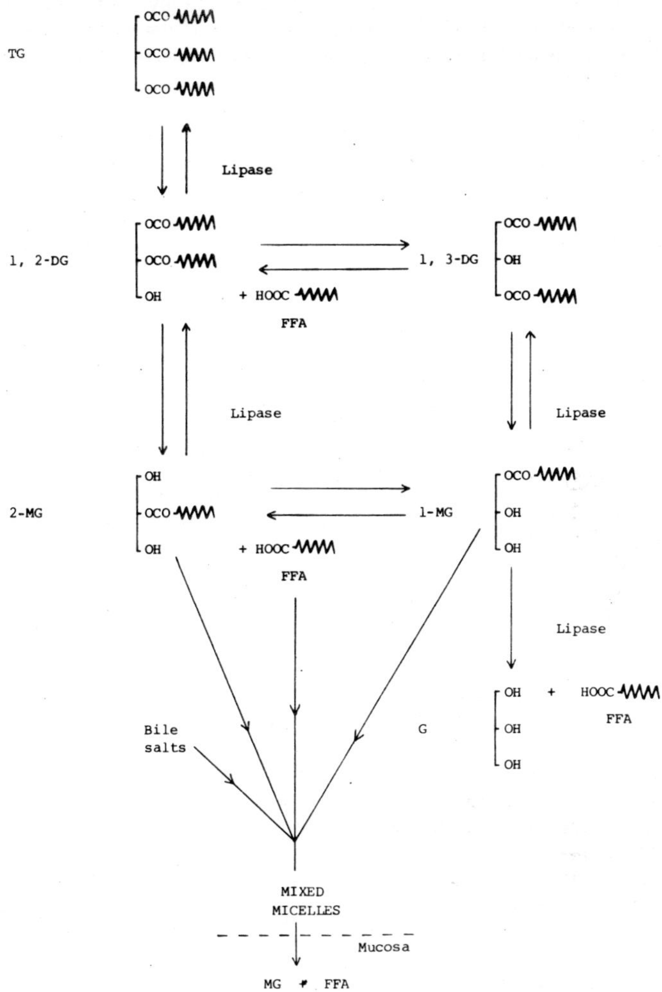

Fig. 8.1   Hydrolysis of triglyceride fats

melting point, it is as readily hydrolysed as oleate (Desnuelle & Savary, 1963). Balls, Matlack & Tucker (1937) showed that the effect of temperature on reaction rate was much greater with long-chain than with short-chain triglycerides.

From our point of view, the relevant consideration is the rate of release of different fatty acids from mixed glycerides. Over the range C_{12} to C_{18} this is not significantly affected by degree of saturation or chain length (Savary &

Desnuelle, 1956; Mattson & Beck, 1956). With mixed glycerides containing C_4, C_6 or C_8 acids, there is marked preferential release of the short chain acid which has been demonstrated both with synthetic triglycerides and triglyceride mixtures (Entressangles, Paséro, Savary, Sarda & Desnuelle, 1961; Clément, Clément & Bezard, 1962a, b) and with milk fat (Clément, Clément, Bézard, Di Costanzo & Paris, 1963; Jack, Freeman, Smith & Mickle, 1963; Jensen, Sampugna & Pereira, 1964; Boudreau & De Man, 1965; Sampugna, Carpenter, Marks, Quinn, Pereira & Jensen, 1966). There is also increased release of fatty acids from position 2 and consequently more total hydrolysis to glycerol when short chain acids are present. This is presumably due to more rapid isomerization.

Phospholipids

In addition to the dietary phospholipids, considerable quantities of lecithin enter the small intestine lumen as part of the bile. These are hydrolysed by enzymes listed in the section on pancreatic secretion, namely the pancreatic phospholipase and lipase. The phospholipase can remove the fatty acid in the 2 position of the glycerol (de Haäs, Sarda & Roger, 1965; de Haas et al., 1968), leaving a product called lysolecithin, and this is probably the function of this enzyme. While the lipase can remove the fatty acid from the 1 position of a phospholipid, it is not established whether this occurs to any extent in the lumen of the gut. While enzymes are present (e.g. mucosal alkaline phosphatase) which could further hydrolyse the lysolecithins, it is quite likely that the latter are largely absorbed as such by the mucosa.

Sterol esters

Sterol esters are hydrolysed in the presence of bile salts by the "cholesterol esterase" of the pancreatic juice to free sterols and free fatty acids. This enzyme will hydrolyse esters of sterols and some other secondary alcohols, but only in the presence of bile salts (Swell, Field & Treadwell, 1953). Both sterol esters and free sterols enter the monoglyceride—fatty acid-bile salt micelles in the lumen and are taken up from micellar solution by the mucosa (Borgström, 1967).

Micellar dispersion

Hofmann & Borgström (1962) showed that when long-chain fatty acids, monoglycerides and bile salts were present together in an aqueous medium they formed a micellar solution. A solution of this type is formed when substances with both a polar or hydrophilic portion of the molecule and a non-polar or lipophilic portion are dissolved. The solute forms micelles, or clusters of a few dozen molecules and in these, if the solvent is water, the hydrophilic portions of the molecules will be exposed to the solvent while the lipophilic portion will be aligned towards the centre. Such micelles can take up appreciable numbers of relatively non-polar molecules, provided that the main components have a suitable combination of water-affinity and fat-affinity. The bile salts are flat with lipophilic hydroxyl groups and amino acid conjugates on one face while the other face is that of a hydrocarbon. The monoglycerides and, to a lesser extent the fatty acids, are hydrophilic at one end, while both are lipophilic at the other. Hofmann & Borgström (1962) found that when bile salts, monoglycerides and fatty acids were together in water, they formed micelles which contained all three solutes, and found that the composition of this micellar solution was similar to that of a filtrate of intestinal contents. They also found that although monoglycerides and fatty acids would form micellar solutions in the presence of bile salts, fatty acids dissolved very slowly unless both bile salts and monoglycerides were present.

Hofmann (1963) showed that monoglycerides with low melting points, such as monoolein, readily formed micellar solutions at 37°C, unlike those with a higher melting point like monopalmitin and monomyristin, but that these higher-melting point monoglycerides dissolved more readily in the presence of those of lower melting point. Savary (1966) showed the same phenomenon with long chain fatty acids in bile salt solutions.

Freeman, Annison, Noakes & Hill (1967) found that the micellar lipid in the small intestine lumen of the pig contained approximately 45% fatty acid, 25% monoglyceride, 10% diglyceride and 5% each of triglyceride, cholesterol, cholesterol esters and phospholipid. These figures are similar to those reported for man and the rat.

Lysolecithin can fulfil the same role as monoglyceride as an aid to micelle formation, and does so in the sheep (Leat, 1965; Harrison & Leat, 1972, 1975). Freeman (1969) showed that lysolecithin was more effective in vitro than monoglyceride in this role, but assumed that owing to its much lower concentration in the lumen of the pig small intestine, it had only a minor role in the pig. He also found that pH was important, micellar solubilization of long chain fatty acids increasing with pH above 6.0.

MUCOSAL UPTAKE

The micelles have an estimated particle size of 30–40 A, small enough to pass between the microvilli of mucosal cells (Borgström, 1967). Monoglycerides and fatty acids are rapidly taken up by the small intestine mucosa from a micellar solution (Hofmann & Borgström, 1964; Johnston & Borgström, 1964). Freeman, Noakes, Annison & Hill (1968) found that long-chain fatty acids of different chain length (C_{12} to C_{18}) were taken up from mixed micellar solutions by pig jejunal slices in vitro at rates related to their micellar concentrations, as were acids of different degrees of saturation (oleic and stearic).

MUCOSAL FAT TRANSPORT AND RESYNTHESIS

After absorption by the mucosa, the long-chain monoglycerides are re-esterified with the long chain fatty acids to diglycerides, which are then further esterified to triglycerides. At the same time, phospholipids are synthesised partly from absorbed and partly from endogenous material. Lipid droplets enclosed in a lipoprotein envelope are formed, apparently in the Golgi apparatus, and travel down ducts in the mucosal cells, to the intercellular spaces. These droplets, called "chylomicrons," then appear in the intestinal lymph, and are the form in which most of the ingested fat reaches the liver or other tissues. They appear in the circulating blood after a fatty meal.

The short-chain fatty acids have a different entry route and pass into the portal blood (Bloom, Chaikoff & Reinhardt, 1951; Greenberger, Rodgers & Isselbacher, 1966), where they appear unesterified (Borgström, 1955). Vodovar, Flanzy & Francois (1967) showed that in pig mucosal cells, the long-chain fatty acids (C_{16} and above) went in combined form into the intercellular spaces leading to the lymph ducts, while with shorter-chain acids, a proportion, increasing with the shortness of the chain, could be detected in the cell cytoplasm and appeared at the base of the cell. Assuming that the process is the same as in other species, the short-chain fatty acids then pass into the capillaries of the portal blood system. The synthetic reactions involved in mucosal fat metabolism were elucidated in laboratory rodents (Brown & Johnston, 1964; Ailhaud, Samuel, Lazdunsky & Desnuelle, 1964) but homogenates of pig mucosa have been shown to carry them out when supplied with the appropriate raw materials and co-factors (Buell & Reiser, 1959; Bickerstaffe & Annison, 1969).

An enzyme capable of breaking down monoglyceride to fatty acid and glycerol is present at a high level in pig small intestine mucosa (Di Nella, Meng & Park, 1960; Bickerstaffe & Annison, 1969), but this hydrolysis only occurs to any degree if monoglyceride is present in excess of the amount of fatty acid required for its esterification (Clément & Paris, 1964; Paris & Clément, 1968).

This combination of reactions identified in the mucosal cell would make it possible for the mucosa to produce triglyceride and phospholipid in the required proportions, starting from fatty acid or monoglyceride or a mixture of them in any proportions. The process of fat transport from the lumen through the mucosa is illustrated diagrammatically in Figs. 8.2 and 8.3.

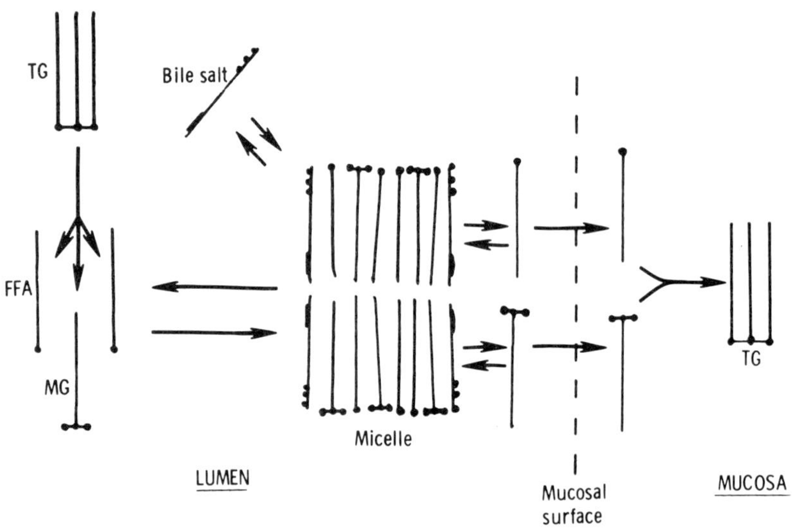

Fig. 8.2 • The hydrolysis, micellar solubilization, absorption and resynthesis of triglycerides in the small intestine.

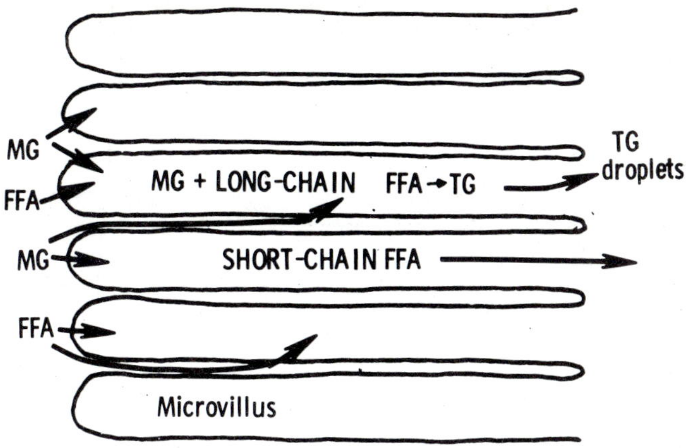

Fig. 8.3 The absorption of fat hydrolysis products by the microvilli of the small intestine mucosa.

The passage of lipids through the mucosa into the intestinal lymph has been demonstrated by a series of electron microscopic studies. The first appearance as extremely small droplets of fat, just within the tips, and to a lesser extent, along the sides of the microvilli of the mucosal cells was shown in rat mucosa by Shiner (1966). The rest of the passage through the mucosa has been demonstrated with pig small intestine by Vodovar and co-workers. If the mucosa had been collected after the pig had had a fatty meal there were many small fat droplets in the "reticular web", a fine network of vessels just below the surface at the base of the microvilli. These fine droplets were most numerous just below the surface in the apical region between the bases of the microvilli and were present deeper in the cell, being progressively larger further into the cell until the Golgi apparatus where they were numerous and large. In still more basal regions of the cell, the droplets were present in intracellular ducts and were similar to those in the Golgi apparatus (Vodovar & Flanzy, 1966, 1967; Vodovar, Flanzy & Francois, 1966, 1967, 1969: Vodovar, Massicard & Flanzy, 1968). The transport of absorbed fat through the mucosal cell is illustrated diagrammatically in Fig. 8.4.

Vodovar, Flanzy & Francois (1966) assumed that esterification or re-esterification produced very small triglyceride droplets at the base of the intervillar spaces and that they then passed along minute ducts to the Golgi apparatus, where they were incorporated into chylomicrons, which then pass through larger ducts to the intercellular spaces and so into the intestinal lymph. The observations of Shiner (1966) make it clear that minute triglyceride droplets are initially formed even earlier, in the microvilli themselves.

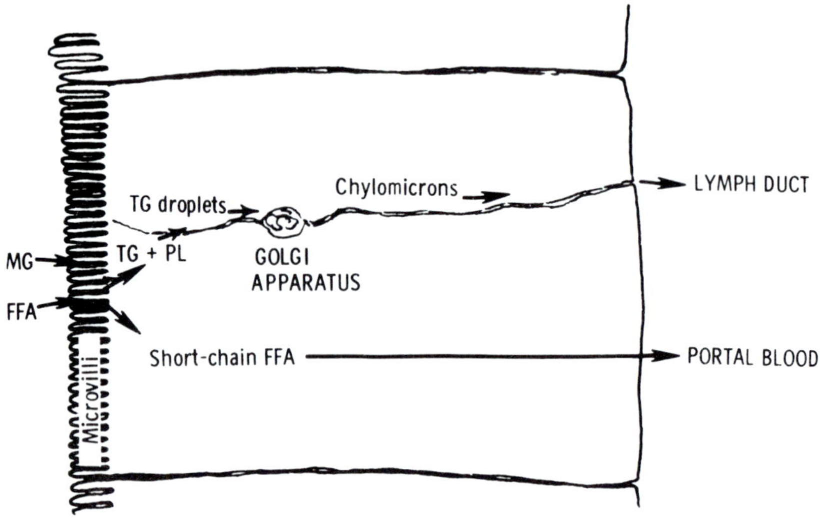

Fig. 8.4 The transport of lipids through the mucosal cells of the small intestine.

LOCATION OF FAT ABSORPTION ALONG THE INTESTINE

Vodovar, Thieulin, Pihet & Flanzy (1965) showed that fat was absorbed along the entire small intestine, but not uniformly. Absorption was slight in the duodenum, but was very intense from a point just distal to the duodenum to about mid-way along the small intestine and then gradually diminished distally. This was the case both with natural triglycerides which were well absorbed (Fig. 8.5), with synthetic triglycerides, some of which were poorly absorbed (Fig. 8.6) and with fatty acids (Fig. 8.7).

QUANTITATIVE ASPECTS OF FAT DIGESTION

Meaningful studies on the utilization of fat will depend on the use of accurate techniques for the measurement of digestibility of fats and of their component fatty acids and on detailed studies of the different stages of the digestive process.

In pig faeces, a major part of the total lipid content is present in the form of calcium soaps. Simple extraction with fat solvents will not remove these soaps and digestibility determined from such an extract would be spuriously high because of inadequate extraction of undigested fatty acids. Toullec, Flanzy & Rigaud (1968) have emphasized the importance of using techniques which extract the fatty acids present in faeces as soaps, preferably distinguishing these from fatty acids present in neutral fats and in the free form. Other, earlier methods, are also suitable if they include an acid hydrolysis stage.

In the studies on fat digestion in pigs in the 1950's and the early and mid 1960's the composition of faecal fat was taken as representing the end point of fat digestion. However, it was shown by Carlson & Bayley (1968) that, in calcul-

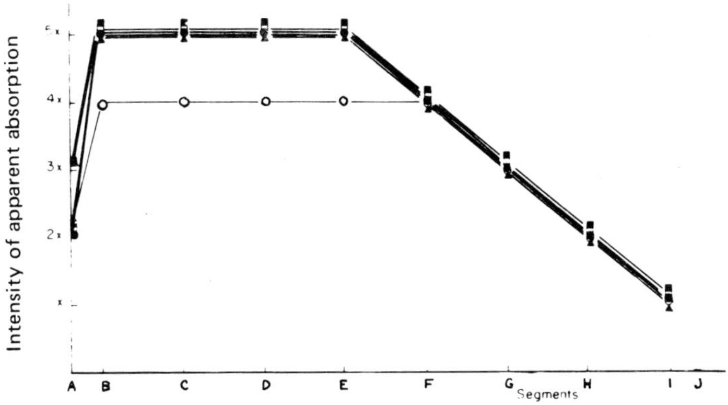

• Coconut oil ○ olive oil □ Cottonseed oil ■ Maize oil ▲ Lard

Fig. 8.5 Absorption of natural triglycerides in the small intestine of the pig. (from Vodovar, Thieulin, Pihet & Flanzy, 1965)

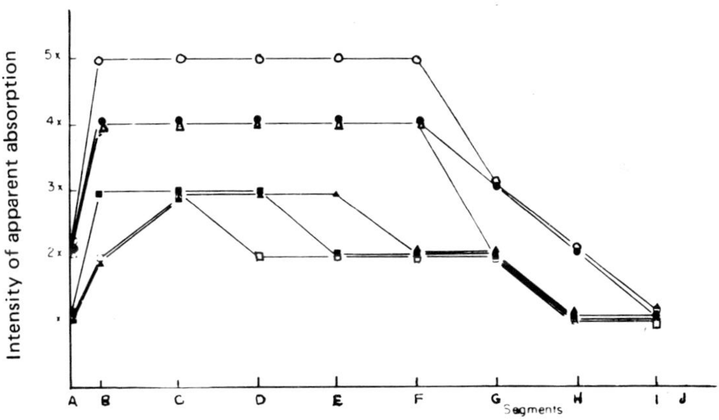

△ Tributyrin • Trilaurin ▲ Tripalmitin ■ Tripalmitin (emulsion)
□ Tristearin ○ Triolein

Fig. 8.6 Absorption of homogeneous triglycerides in the small intestine of the pig. (from Vodovar, Thieulin, Pihet & Flanzy, 1965)

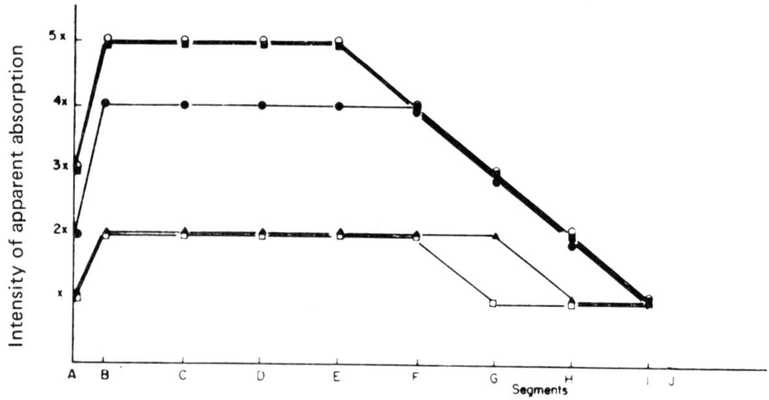

● Lauric acid ▲ Palmitic acid □ Stearic acid ○ Oleic acid ■ Linoleic acid

Fig. 8.7 Absorption of free fatty acids in the small intestine of the pig. (from Vodovar, Thieulin, Pihet & Flanzy, 1965)

ating apparent digestibilities of individual fatty acids from inputs in feed and outputs in faeces, workers had been failing to recognise the effects of microflora in the caecum and lower gut, especially the hydrogenation of individual unsaturated fatty acids. Carlson & Bayley (1968) compared the fatty acid compositions of the fat in the digestive systems of a germ-free pig and of a conventionally-reared pig, both fed on the same diet. In the germ-free pig, stearic acid levels in the lower gut were relatively low whereas, in the conventional pig, stearic acid predominated in the fat present in the lower gut. It was presumed that hydrogenation of unsaturated C_{18} fatty acids was responsible for the difference observed and this conversion will be likely to occur in all conventional pigs, thus making digestibility coefficients for individual C_{18} fatty acids very unreliable.

Lack of a fat-soluble, indigestible marker substance which would move down the gut at the same speed as the fat with which it was incorporated, also hindered advances in the study of the process of fat digestion. Chromic oxide, used by many workers as a marker, was found by Carlson & Bayley (1968) to separate from the fat in the digestive system, invalidating its use as a marker for studies of the process of digestion of fat. Carlson & Bayley (1970) suggested the use of tridodecyl glyceryl ether as a fat absorption indicator and this substance was used in the studies of Swiss, Carlson & Bayley (1971). The latter confirmed that unsaturated C_{18} fatty acids are hydrogenated during their passage through the lower gut, giving rise to spuriously high 'indigestible' stearic acid figures. Freeman, Holme & Annison (1968) also obtained evidence of extensive hydrogenation of oleic acid in the digestive tract of growing pigs and, in the sow, Salmon–Legagneur & Friend (1971) found indications of hydrogenation of unsaturated C_{18} fatty acids to stearic acid in the caecum and colon. In addition to the formation of stearic acid in the lower gut, fatty acids with odd numbers of carbon atoms e.g. pentadecanoic and heptadecanoic acids are formed (Bayley &

Lewis, 1965a), and these are presumed to be the result of bacterial action within the gut. The part played by bacteria in lowering the proportion of unsaturated fatty acids in pig faeces and raising the proportion of saturated acids was shown strikingly by Mason & Just (1976) who fractionated the fatty acids in the food, jejunal contents, ileal contents and faeces of 90 kg pigs on two diets with or without antibiotics. On the maize starch diet the faecal tri-unsaturated C_{18} acids were 0.7% without antibiotic in the diet and 3.8% with, while the C_{18} saturated acids were 41.7% without antibiotic and only 5.1% with. On the potato starch diet the faecal tri-unsaturated C_{18} acids were 0.8% without antibiotic and 3.1% with, and the saturated C_{18} acids were 44.1% without antibiotic and only 6.5% with antibiotic in the diet. The fatty acid composition in the jejunum and ileum was unaffected by the presence or absence of antibiotic. Clearly, in the absence of antibiotic the bacteria in the large intestines had almost completely eliminated the highly unsaturated acids and were responsible for the presence of most of the saturated C_{18} (stearic) acid.

Developments in the technique of thin layer chromatography have made it possible to separate the main lipid fractions present in gut contents with relative ease, compared with the earlier use of column chromatography. After separation of components in this way, the fatty acid composition of the individual fractions can be determined by gas chromatography so that the components of triglycerides can be followed through the digestive process as food passes along the gut (Kidder & Manners, 1972). Now that appropriate techniques are available, the study of lipid digestion could become more precise. However, progress is likely to be limited by the very high labour requirement due to the multiplicity of lipid fractions to be monitored if such investigations are to be comprehensive, and very little work of this type has been reported.

The measurement of fat digestibility is also complicated by the entry of appreciable quantities of endogenous fat into the small intestine as a component of the lecithin in the bile. This not only influences the measurement of the overall digestibility but also that of individual fatty acids, as the fatty acid composition of the endogenous fat may differ materially from that in the diet.

While most of the published work on fat digestibility in the pig consists of measurement of apparent digestibility, some measurements have included corrections for endogenous fat, and others, using isotope techniques, have measured true digestibilities directly.

As is described later, digestibility of fats alters with age. In addition, it varies with the nature of the fat, diminishing with increasing chain length of the fatty acid and increasing with the degree of unsaturation. The relationship between apparent digestibility of fats and chain length was shown in piglets up to 4 weeks by Braude & Newport (1973), in piglets of 3 and 7 weeks by Lloyd and Crampton (1957) and Lloyd, Crampton & MacKay, (1957), in 3 month old pigs by Fritz (1962), in 16–18 kg pigs by Freeman, Holme and Annison (1968) and in 27 kg pigs by Bayley & Lewis (1965b). The effect of unsaturation was also shown in the experiments of Fritz (1962), Bayley & Lewis (1965b) and Freeman, Holme & Annison (1968) and to a slight extent in those of Lloyd & Crampton (1957).

True digestibilities were calculated for the various fatty acids under their experimental conditions by Bayley and Lewis (1965b) who determined the endogenous fat on a fat-free diet and assumed that the same value applied on other diets. Freeman, Holme & Annison (1968) using labelled fats reported that the endogenous fat varied with the diet from 5 g to over 32 g per day on 16–18 kg pigs. However, their findings on corrected digestibilities do not differ substantially from those of Bayley & Lewis (1965b), who found that palmitic acid was 64% digested in a tripalmitin diet and 68–91% digested when fed in diets with other fats of higher digestibility. For stearic acid, values of 40–50% were found in most diets by both sets of workers, and for oleic acid 90–100%.

The wide variation in digestibility of an individual acid with diets is due to the marked effect a more digestible fat has in increasing the digestibility of a less digestible one. There is also the effect noted by both the above authors, of the reduced digestibility of more digestible acids like oleic, when fed in diets rich in saturated acids like stearic. This is presumably due to the effect, already referred to, in which the micellar dispersion of high melting-point fatty acids or monoglycerides, is improved by the presence of low melting point fatty acids or monoglycerides. In consequence, oleic and stearic acids, or lauric, myristic, palmitic and stearic, are taken up by the pig intestine wall at equal rates from a mixed micellar solution perfused into the lumen (Freeman, Noakes, Annison & Hill, 1968). In addition, the position of the fatty acid in the glyceride molecule affects its digestibility. Freeman, Holme & Annison (1968) showed differences in digestibilities for individual fatty acids between natural fats and the same fats after inter-esterification.

The overall effect of these factors was shown by the work of Vodovar, Thieulin, Pihet & Flanzy (1965), already referred to (Figs. 8.5, 8.6 & 8.7) who found that a range of different natural fats all had similar and high digestibilities but that synthetic triglycerides of single fatty acids, or free single fatty acids, had digestibilities which varied with chain length and degree of saturation.

Digestion of fat by suckled pigs

The sow-reared pig receives a diet which is very rich in fat (42% of the dry matter; Braude, 1964). The fat provides approximately 60% of the dietary energy and so efficient digestion and absorption of fats is necessary. Barber, Braude & Mitchell (1955) found that 0.81 g of milk dry matter was needed per g of weight gain by sow-reared piglets. Thus, the dry matter of sows' milk is converted into weight gain very efficiently, indicating that the sow-reared piglet has little or no difficulty in digesting and utilising large amounts of fat. Frobish, Hays, Speer & Ewan (1967) reported the mean digestibility coefficient for sows' milk fat to be 95.1%, using 2-day-old pigs. However, it should be remembered that the suckled pig receives its food in evenly-spaced meals approximately hourly throughout the day and night and that the lipid in sows' milk is in a very finely divided state (Whittlestone, 1952) compared with, for example, that in cows' milk (Whittlestone, 1954). Thus the high intake of fat is in a form presenting optimum opportunity for digestion and absorption.

Fat in early weaning

The question of the optimum level, type and form of presentation of fat in diets for early-weaned pigs is by no means resolved at the present time. Fat digestion has proved difficult to study because of the variations in composition of fats and the difficulties in monitoring all the various lipid fractions during the passage of food through the gut. As shown earlier, it is only in fairly recent years that suitably precise techniques have become available to enable one to envisage a breakthrough in knowledge in this field.

Diets used by early workers studying the nutritional requirements of artificially reared piglets often contained large amounts of fat, the highest level used being that in the experiments of Reber, Whitehair & MacVicar (1953), where 35–49% lard was added. These early diets appeared to give satisfactory results. Slightly later, conflicting results began to appear about the value of added fat in artificial diets. The position was complicated by the differences between the experiments that were reported. Weaning age varied from 2 days of age to several weeks of age. In some experiments diets were fed dry and in others in liquid form. The raised energy levels in high fat diets were not always accompanied by raised protein levels to keep the energy/protein ratio constant. In some cases fats were emulsified into liquid diets and in others they were merely added in liquid form (after melting if necessary) to the dry ingredients in a food mixer. The results of the early work were reviewed by Manners & McCrea (1958), Lucas & Lodge (1961) and Braude (1964).

In interpreting the results of different experiments it should be remembered that with increasing age there is an improvement in the ability of pigs to digest fat in artificial diets (first demonstrated by Cunningham & Brisson, 1955). As mentioned earlier, Lloyd & Crampton (1957), Lloyd, Crampton & MacKay (1957) and Braude & Newport (1973) showed that digestibility of fats by early-weaned pigs was inversely related to fatty acid chain length. In the studies of Lloyd & Crampton there seemed to be no relationship between the degree of saturation or unsaturation of the fats and their digestibility by piglets, but Braude and Newport (1973) found a higher digestibility for unsaturated fatty acids, and a low digestibility for beef tallow as compared with more unsaturated fats (butterfat, coconut oil, soya bean oil). Lloyd, Crampton & MacKay (1957) confirmed that digestibility of fats by piglets improved with age, and showed that the improvement with age was greater for those fats which were poorly digested in the young piglet (i.e. those of long fatty acid chain length).

It has been pointed out by Hamilton and McDonald (1969) that the method of extraction of faecal lipids used by Lloyd, Crampton & MacKay (1957) (a single neutral diethyl ether extraction) would only have removed a portion of the faecal soaps (see earlier) and thus the coefficients of apparent digestibility calculated by Lloyd et al. (1957) are likely to have been higher than if a more efficient extraction procedure had been followed.

Manners & McCrea (1963), adding fat to diets for 2-day weaned piglets were unable to get worthwhile responses from the addition of high levels of fats. They did not measure digestibility of fats, but economy of food conversion was not improved in the diets with added fat and thus these diets were used less efficiently than those with low levels of fat.

The reason why the sow-reared pig can utilise large amounts of fat while the artificially-reared pig can only apparently utilise limited amounts may be related to a variety of factors. In the first place, artificially-reared piglets are usually fed much less frequently but with greater amounts per feed than sow reared piglets, with the consequent risk of overloading the digestive system. Also, fat is unlikely to be presented in such a finely divided state in artificial diets as it is in sows' milk. The fatty acid composition of artificial diets will vary and will differ from that in sows' milk. Pancreatic lipase levels may be lower in artificially reared piglets than in sow-reared animals (as in the experiments of Hartman, Hays, Baker, Neagle & Catron, 1961). The pH in the small intestine in artificially-reared pigs may be such that micellar solubilization of fatty acids may be limited. In the small intestine of 2- and 3-week old pigs the mean pH has been found to be below 6.0 at most sites (Manners, 1970). Artificial diets need to be rich in calcium to support good bone development in the growing piglet and the high levels of calcium present in small intestine contents of piglets on such diets may depress rates of micelle formation. Finally, it might be worth while to investigate the effects of inclusion of a variety of amphiphilic substances in high fat diets for artificial rearing to see if the efficiency of digestion of fat could be improved by such additions.

REFERENCES

Ailhaud, C., Samuel, D., Lazdunski, M. & Desnuelle, P. (1964) Some observations on the mode of action of mono- and diglyceride transacylase of intestinal mucosa. *Biochim. biophys. Acta, 84,* 643–664.

Balls, A.K., & Matlack, M.B. (1938) Mode of action of pancreatic lipase. *J. biol. Chem., 123,* 679–686.

Balls, A.K., Matlack, M.B. and Tucker, I.W. (1937) The hydrolysis of glycerides by crude pancreatic lipase. *J. biol. Chem., 122,* 125–127.

Barber, R.S., Braude, R. & Mitchell, K.G. (1955) Studies on milk production of Large White pigs. *J. agric. Sci., Camb., 46,* 97–118.

Bayley, H.S. & Lewis, D. (1965a) The fats in pig feeding. I. Pig faecal fat not of immediate dietary origin. *J. agric. Sci. Camb., 64,* 367–372.

Bayley, H.S. & Lewis, D. (1965b) The use of fats in pig feeding. II. The digestibility of various fats and fatty-acids. *J. agric. Sci. Camb., 64,* 373–378.

Benzonana, G., Entressangles, B., Marchis-Mouren, G., Paséro, L., Sarda, L. & Desnuelle, P. (1964) Further studies on pig pancreatic lipase. *in* Metabolism and physiological significance of lipids. Eds. Dawson, R.M.C. & Rhodes, D.N. pp. 141–154. John Wiley & Sons, N.Y.

Bickerstaffe, R. & Annison, E.F. (1969) Triglyceride synthesis by small-intestinal epithelium of the pig, sheep and chicken. *Biochem. J. 111,* 419–429.

Bloom, B., Chaikoff, I.L. & Reinhardt, W.O. (1951) Intestinal lymph as pathway for transport of absorbed fatty acids of different chain lengths. *Am. J. Physiol., 166,* 451–455.

Borgström, B. (1954a) Effect of taurocholic acid on the pH/activity curve of rat pancreatic lipase. *Biochim. biophys. Acta, 13,* 149–150.

Borgström, B. (1954b) On the mechanism of pancreatic lipolysis of glycerides. *Biochim. biophys. Acta, 13,* 491–504.

Borgström, B. (1955) Transport of ^{14}C-decanoic acid in portal and inferior vena cava blood during absorption in the rat. *Acta physiol. scand., 34,* 71–74.

Borgström, B. (1964) Influence of bile salt, pH and time on the action of pancreatic lipase; physiological implications. *J. Lipid Res., 5,* 522–531.

Borgström, B. (1967) Absorption of fats. *Proc. Nutr. Soc., 26,* 34–46.

Borgström, B. (1975) On the interaction between pancreatic lipase and colipase and the substrate, and the importance of bile salt. *J. Lipid Res., 16,* 411–417.

Borgström, B. & Donnér, J. (1975) Binding of bile salts to pancreatic colipase and lipase. *J. Lipid. Res., 16,* 287–292.

Borgström, B. & Donnér, J. (1976) On binding of bile salt to pancreatic lipase. *Biochem. biophys Acta., 450,* 352–357.

Boudreau, A. & De Man, J.M. (1965) Mode of action of pancreatic lipase on milk fat. *Can. J. Biochem., 43,* 1799–1805.

Braude, R. (1964) Problems in pig rearing. European association for animal production. Commission of pig production. Göttingen.
Braude, R. & Newport, M.J. (1973) Artificial rearing of pigs. 4. The replacement of butterfat in a whole-milk diet by either beef tallow, coconut oil or soya bean oil. *Br. J. Nutr., 29,* 447–455.
Brown, J.L. & Johnston, J.M. (1964) Utilization of 1- and 2-monoglycerides for intestinal triglyceride synthesis. *Biochim. biophys. Acta, 84,* 448–547.
Buell, G.C. & Reiser, R. (1959) Glyceride glycerol precursors in the intestinal mucosa. *J. biol. Chem. 234,* 217–219.
Carlson, W.E. & Bayley, H.S. (1968) Utilization of fat by young pigs: fatty acid composition of ingesta in different regions of the digestive tract and apparent and corrected digestibilities of corn oil, lard and tallow. *Can. J. Anim. Sci., 48,* 315–322.
Carlson, W.E. & Bayley, H.S. (1970) Tridodecyl glycerol ether in fat absorption studies. *J. Anim. Sci., 31,* 1019.
Clément, G., Clément, J. & Bézard, J. (1962a) Action de la lipase pancréatique humaine sur les triglycerides mixtes, symmetrique de synthèse comportant des acides gras longs et de l'acide butyrique. *Arch. Sci. Physiol. 16,* 213–225.
Clément, G. Clément, J. & Bézard, J. (1962b) Action of human pancreatic lipase on synthetic mixed symmetrical triglycerides of long-chain acids and butyric acid. *Biochim. biophys. Res. Commun., 8,* 238–242.
Clément, G., Clément, J. Bézard, J., Di Costanzo, G. & Paris, R. (1963) Hydrolysis of the triglycerides of butter by pancreatic lipase. Location of butyric acid. *J. Dairy Sci., 46,* 1423–1425.
Clément, G. & Paris, R. (1964) Absorption d'acides gras et de monoglycérides marquées par un segment d'intestin de rat in situ. *C.r. hebd. Séanc. Acad. Sci., Paris, 258,* 6006–6008.
Cunningham, H.M. & Brisson, G.J. (1955) The utilization of lard by baby pigs. *Can. J. agric. Sci., 35,* 371–376.
Dawson, A.M. (1971) The absorption of fat. *J. clin. Path., 24,* Suppl. (Roy. Coll. Path.), *5,* 77–84.
Desnuelle, P. & Savary, P. (1963) Specificities of lipases. *J. Lipid Res. 4,* 369–384.
Di Nella, R.R., Meng, H.C. & Park, C.R. (1960) Properties of intestinal lipase. *J. biol. Chem., 235,* 3076–3081.
Entressangles, B., Paséro, L., Savary, P., Sarda, L. & Desnuelle, P. (1961) Influence de la nature des chaines sur la vitesse de leur hydrolyse par la lipase pancréatique. *Bull. Soc. chim. Biol., 43,* 581–591.
Entressangles, B., Sari, H. & Desnuelle, P. (1966) On positional specificity of pancreatic lipase. *Biochim. biophys. Acta, 125,* 597–600.
Erlanson, C., Fernlund, P. & Borgström, B. (1973) Purification and characterization of two proteins with co-lipase activity from porcine pancreas. *Biochim. biophys. Acta, 310,* 437–445.
Freeman, C.P. (1969) Properties of fatty acids in dispersions of emulsified lipid and bile salt and the significance of these properties in fat absorption in the pig and the sheep. *Br. J. Nutr., 23,* 249–263.
Freeman, C.P., Annison, E.F., Noanes, D.E. & Hill, K.J. (1967) The absorption of micellar fat in pigs. *Proc. Nutr. Soc., 26,* vii.
Freeman, C.P., Holme, D.W. & Annison, E.F. (1968) The determination of the true digestibilities of interesterified fats in young pigs. *Brit. J. Nutr., 22,* 651–660.
Freeman, C.P., Noakes, D.E., Annison, E.F. & Hill, K.J. (1968) Quantitative aspects of intestinal fat absorption in young pigs. *Br. J. Nutr., 22,* 739–749.
Fritz, Z. (1962) Strawnośc tluszczu surowego i wyzszuch kwasów tluszczowych u świń oznaczona metoda wskaznikowa z zastosowaniem chromatografii bibulowej. *Roczn. Nauk Roln., 81,* 23–42.
Frobish, L.T., Hays, V.W., Speer, V.C. & Ewan, R.C. (1967) Digestion of sow milk fat and effect of diet form on fat utilization. *J. Anim. Sci., 26,* 1478–1479.
Greenberger, N.J., Rodgers, J.B. & Isselbacher, K.J. (1966) Absorption of medium and long chain triglycerides. Factors influencing their hydrolysis and transport. *J. clin. Invest., 45,* 217–227.
de Haas, G.H., Postema, N.M., Nieuwenhuizen, W. & van Deenen, L.L.M. (1968) Purification and properties of phospholipase A from porcine pancreas. *Biochim. biophys. Acta., 159,* 103–117.

de Haas, G.H., Sarda, L. & Roger, J. (1965) Positional specific hydrolysis of phospholipids by pancreatic lipase. *Biochim. biophys. Acta, 106,* 638–640.
Hamilton, R.M.G. & McDonald, B.E. (1969) Effect of dietary fat source on the apparent digestibility of fat and the composition of fecal lipids in the young pig. *J. Nutr., 97,* 33–41.
Harrison, F.A. & Leat, W.M.F. (1972) Absorption of palmitic, stearic and oleic acids in the sheep in the presence or absence of bile and/or pancreatic juice. *J. Physiol., Lond., 225,* 565–576.
Harrison, F.A. & Leat, W.M.F. (1975) Digestion and absorption of lipids in non-ruminant and ruminant animals: a comparison. *Proc. Nutr. Soc., 34,* 203–210.
Hartman, D.A., Hays, V.W., Baker, R.O., Neagle, L.H. & Catron, D.V. (1961) Digestive enzyme development in the young pig. *J. Anim. Sci., 20,* 114–123.
Hofmann, A.F. (1963) The behaviour and solubility of monoglycerides in dilute, micellar bile salt solution. *Biochim. biophys. Acta, 70,* 306–316.
Hofmann, A.F. & Borgström, B. (1962) Physico – chemical state of lipids in intestinal content during their digestion and absorption. *Fedn. Proc. Fedn Am. Socs. exp. Biol., 21,* 43–50.
Hofmann, A.F. & Borgström, B. (1964) The intraluminal phase of fat digestion in man: the lipid content of the micellar and oil phases of intestinal content obtained during fat digestion and absorption. *J. clin. Invest., 43,* 247–257.
Jack, E.L., Freeman, C.P., Smith, L.M. & Mickle, J.B. (1963) Pancreatic lipase hydrolysis of cow milk fat. *J. Dairy Sci., 46,* 284–290.
Jensen, R.C., Sampugna, J. & Pereira, R.L. (1964) Pancreatic lipase. Lipolysis of synthetic triglycerides containing a trans fatty acid. *Biochim. biophys. Acta, 84,* 481–483.
Johnston, J.M. (1968) Mechanism of fat absorption. *in* Handbook of Physiology, Sect. 6. Alimentary canal Vol. III. Intestinal absorption Ch. 70 pp 1353–1376. Williams & Wilkins, Baltimore.
Johnston, J.M. & Borgström, B. (1964) The intestinal absorption and metabolism of micellar solutions of lipids. *Biochim. biophys. Acta, 84,* 412–423.
Kidder, D.E. & Manners, M.J. (1972) The development of digestive function in the piglet. *Vet. Annual, 13,* 57–71.
Leat, W.M.F. (1965) Possible function of bile and pancreatic juice in fat absorption in the ruminant. *Biochem. J., 94,* 21P.
Lloyd, L.E. & Crampton, E.W. (1957) The relation between certain characteristics of fats and oils and their apparent digestibility by young pigs, young guinea-pigs and pups. *J. Anim. Sci., 16,* 377–382.
Lloyd, L.E., Crampton, E.W. & MacKay, V.G. (1957) The digestibility of ration nutrients by three- versus seven-week old pigs. *J. Anim. Sci., 16,* 383–388.
Lucas, I.A.M. & Lodge, G.A. (1961) The nutrition of the young pig. Tech. Comm. No. 22. Commonwealth Bureau of Animal Nutrition, Aberdeen, Scotland.
Manners, M.J. (1970) Milk replacers for piglets. *J. Sci. Fd. Agric., 21,* 333–340.
Manners, M.J. & McCrea, M.R. (1958) The nutrition of the baby pig. *Agric. Rev., 3,* (10), 37–45.
Manners, M. J. & McCrea, M.R. (1963) Protein requirements of baby pigs. 2. Requirement on high-fat diets and effect of replacing carbohydrate by fat. *Brit. J. Nutr., 17,* 357–372.
Mason, V.C. & Just, A. (1976) Bacterial activity in the hind-gut of pigs. 1. Its influence on the apparent digestibility of dietary energy and fat. *Z. Tierphysiol. Tierernähr. Futtermittelk, 36,* 301–310.
Mattson, F.H. & Beck, L.W. (1956) Specificity of pancreatic lipases for the primary hydroxyl groups of glycerides. *J. biol. Chem., 219,* 735–740.
Mattson, F.H. & Volpenhein, R.A. (1964) The digestion and absorption of triglycerides. *J. biol. Chem., 239,* 2772–2777.
Mattson, F.H. & Volpenhein, R.A. (1972) Hydrolysis of fully esterified alcohols containing from one to eight hydroxyl groups by the lipolytic enzymes of rat pancreatic lipase. *J. Lipid Res., 13,* 325–328.
Momsen, W.E. & Brockman, H.L. (1976a) Effects of colipase and taurocholate on the catalytic and physical properties of pancreatic lipase B at an oil-water interface. *J. biol. Chem., 251,* 378–383.
Momsen, W.E. & Brockman, H.L. (1976b) Inhibition of pancreatic lipase B activity by taurocholate and its reversal by colipase. Mechanism of action. *J. biol. Chem., 251,* 384–388.

Morgan, R.G.H., Barrowman, J., Filipek-Wender, H. & Borgström, B. (1968) The lipolytic enzymes of rat pancreatic juice. *Biochim. biophys. Acta, 167,* 355–366.
Paris, R. & Clément, G. (1968) Biosynthèse de triglycérides a partir de 20- monopalmitine doublement marquée dans la muqueuse intestinale de rat. *Biochim. biophys. Acta, 152,* 63–74.
Raghavan, S.S. & Ganguly, J. (1969) Studies on the positional integrity of glyceride fatty acids during digestion and absorption in rats. *Biochem. J., 113,* 81–87.
Reber, E.F., Whitehair, C.K. & MacVicar, R. (1953) The effect of level of protein fed baby pigs. *J. Nutr., 50,* 451–458.
Salmon-Legagneur, E. & Friend, D.W. (1971) Digestion des graisses chez la truie. *Annls Biol. anim. Biochim. Biophys., 11,* 312–314.
Sampugna, J., Carpenter, D.L., Marks, T.A., Quinn, J.G., Pereira, R.L. & Jensen, R.G. (1966) Interpretation of pancreatic lipase analysis of cow's milk fat triglyceride structure. *J. Dairy Sci., 49,* 163–168.
Sarda, L. & Desnuelle, P. (1958) Action de la lipase pancréatique sur les esters en emulsion. *Biochim. biophys. Acta, 30,* 513–521.
Savary, P. (1966) Sur la solubilisation micellaire des acids palmitique, stearique et oléique dans les solutions d'acides biliares conjugués. *Biochim. biophys. Acta, 125,* 328–334.
Savary, P. & Desnuelle, P. (1956) Sur quelques elements de specificité pendant l'hydrolyse enzymatique des triglycerides. *Biochim. biophys. Acta, 21,* 349–360.
Schonheyder, F. & Volqvartz, K., 1954 Studies on the lipolytic enzyme action. VI. Hydrolysis of trilauryl glycerol by pancreatic lipase. *Biochim. biophys. Acta, 15,* 288–290.
Shiner, M. (1966) Electron-microscope evidence for intramicrovillous fat absorption by the small intestinal epithelium of rats. *Gut, 7,* 107–113.
Swell, L., Field, H. & Treadwell, C.R. (1953) Role of bile salts in the activity of cholesterol esterase. *Proc. Soc. exp. Biol., N.Y., 84,* 417–420.
Swiss, L.D., Carlson, W.E. & Bayley, H.S. (1971) Influence of level of free fatty acid in dietary fat on fatty acid uptake by piglets. *Fedn Proc. Fedn. Am. Socs. Exp. Biol., 30,* 236.
Toullec, R., Flanzy, J. & Rigaud, J. (1968) Dosage des lipides des fèces. Extraction séparée, importance et composition en acides gras des lipides non-esterifiés et de ceux complexes insoluble. *Annls Biol. anim. Biochim. Biophys., 8,* 281–289.
Vodovar, N. & Flanzy, J. (1966) Mecanisme de la pénétration et voie d'acheminement des graisses dans la cellule absorbante de l'intestin grêle du porc. *Annls Biol. anim. Biochim. Biophys., 6,* 315–332.
Vodovar, N. Flanzy, J. (1967) Voie de pénétration des lipides ingérés dans le stroma des villosités. *C.r. hebd. Seanc. Acad. Sci., Paris, 264,* 850–853.
Vodovar, N., Flanzy, J. & Francois, A.C. (1966) Pénétration et acheminement des graisses dans la cellule épithéliale de l'intestin grêle du porc. *C.r. hebd. Séanc. Acad. Sci., Paris, 262,* 812–815.
Vodovar, N., Flanzy, J. & Francois, A.C. (1967) Aspect morphologique de la répartition chyloportale des acides gras absorbes au niveau de l'intestin grêle. *C.r. hebd. Séanc. Acad. Sci., Paris, 265,* 564–567.
Vodovar, N., Flanzy, J. & Francois, A.C. (1969) Lieu de la formation des chylomicrons; leur importance pour l'absorption normale des acides gras à chaine longue. Aspect morphologique. *Annls Biol. anim. Biochim. Biophys., 9,* 219–232.
Vodovar, N., Massicard, N. & Flanzy, J. (1968) Formation et rôle des chylomicrons au cours de l'absorption des acides gras à chaine longue. *C.R. hebd. Séanc. Acad. Sci., Paris, 266,D* 814–817.
Vodovar, N., Thieulin, C., Pihet, A. & Flanzy, J. (1965) Lieu d'absorption des graisses dans l'intestin grêle du porc. *Annls Biol. anim. Biochim. Biophys, 5,* 249–265.
Whittlestone, W.G. (1952) The distribution of fat globule size in sow's milk. I. The effect of sampling at intervals throughout milking. *J. Dairy Res., 19,* 127–132.
Whittlestone, W.G. (1954) The physical properties of cows' milk as influenced by stage of lactation. The size distribution of butterfat globules. *J. Dairy Res., 21,* 50–54.

CHAPTER 9
DIGESTIBILITY

INTRODUCTION

Since the pig is used to convert a variety of foods into meat for human consumption, the efficiency with which it carries out this conversion is of crucial importance to the existence of the pig industry. For a diet to be useful to the pig, its components must be digestible and so the digestibility of feeds and their constituents is of great economic importance.

The traditional methods of proximate analysis of feedstuffs have provided a vast amount of data on the crude composition of feeds. The components measured are dry matter, ash, organic matter (dry matter minus ash), crude protein (N x 6.25), ether extract, crude fibre and nitrogen-free extractives (N.F.E.), the latter two fractions being intended to represent the relatively indigestible fibrous components and the (sugars + starch), respectively.

Applying the traditional methods of analysis to food and faeces, and measuring inputs and outputs of food and faeces, one can arrive at measures of the extent of digestibility of the components listed above. Generally the digestibility measurements are reported in terms of 'apparent digestibility', since it is difficult to devise appropriate corrections for the amount of digestive secretions and other waste products which are irretrievably mixed up with the undigested feed residues of which the faeces are largely composed. The coefficient of apparent digestibility of dry matter (or, preferably, organic matter) provides a valuable index of the overall digestibility of a diet. A more complete picture of digestibility emerges if one calculates apparent digestion coefficients for crude protein, ether extract, crude fibre and nitrogen-free extractives.

APPARATUS AND TECHNIQUES FOR THE DETERMINATION OF DIGESTIBILITY

Precise determination of the inputs of food and outputs of faeces is difficult in the pig. The feeding of restricted amounts of food will help to ensure that the experimental animal eats up as much as possible of each feed, thus minimizing wastage and the problem of measuring feed refusals. The quantitative collection of faeces necessitates either a harness (difficult to fit and adjust on a streamlined animal like the pig) or a complex digestion crate in which the excreta from the animal are separated and collected via chutes, screens and funnelling arrangements. Animals need to be accustomed to such apparatus and the rapid growth potential of the pig necessitates facilities for continual adjustment of harness (if any) and crate.

Digestibility is usually measured with male animals in a metabolism crate such as that described by Allen, Barber, Braude & Mitchell (1963) (Later modified by Frape, Wolf, Wilkinson & Chubb, 1968). Difficulties arise in keeping faeces entirely separate from urine, and in the design used by Allen et al. (1963), the pig wore a harness to which a light-weight plastic coated "apron" was attached. The purpose of the latter was to deflect faeces backwards and away from the urine-collecting arrangements.

There has been a great deal of interest over the past 20 years or so in the use of indigestible marker substances (particularly chromic oxide) added at a constant low level to the feed. When such a marker is used the percentage apparent digestibility of a nutrient can be calculated from the following formula:–

$$\frac{\text{ratio of marker to nutrient in faeces} \text{ minus } \text{ratio of marker to nutrient in food}}{\text{ratio of marker to nutrient in faeces}} \times \frac{100}{1}$$

(Recently there has been change in the way digestibility is expressed. Instead of a percentage, the current convention is to calculate an index in which unity represents 100%).

The main difficulty with using chromic oxide is its extreme insolubility which, while making it a very suitable indigestible marker, makes it rather difficult to dissolve and estimate chemically for the purpose of calculating ratios of Cr_2O_3 to nutrient and, hence, digestibility. Some workers have claimed practically complete recovery of Cr_2O_3. Others have been unable to recover the material satisfactorily. A large number of variations on the techniques of estimation of Cr_2O_3 have been published over the years and it seems likely, in the authors' opinion, that there is some difficulty in this. Recently, small scale experiments by McCarthy, Aherne & Okai (1973) have indicated that the natural insoluble substances in pig feeds may, in some circumstances, form a satisfactory inert marker. Apparent digestibility coefficients calculated using ash in feed and faeces insoluble in 4N HCl were close to the coefficients determined on the same animals by total faecal collection, and closer than were the coefficients calculated using Cr_2O_3 as a marker. (However, they were unable to recover Cr_2O_3 in faeces quantitatively).

FEEDINGSTUFFS AND THEIR FIBRE CONTENT

The pig is fed largely on concentrated foods and the ingredients of the latter are generally chosen for their ease of digestion. For example, cereal grains predominate in rations used in the United Kingdom. Foods high in fibre are usually diverted to the feeding of ruminants, although adult pigs will happily consume considerable amounts of very fibrous materials (e.g. wheat straw) if given the opportunity. Young pigs, too, consume straw bedding as part of their daily activities when kept in pens where bedding is available. However, fibrous ingredients added to the concentrates fed to pigs can be expected to slow down the rate of growth during the fattening period. In fact, the addition of fibre has been used as a means of restricting growth with the aim of improving carcass quality by reducing fatness of the pig at slaughter weight (e.g. Crampton, Ashton & Lloyd, 1954a) although quantitative restriction of feed (e.g. Crampton, Ashton & Lloyd, 1954b) is a much more common and efficient method of achieving reduced carcass fatness.

Since fibrousness of feedstuffs is important in relation to their feeding value to pigs, the laboratory determination of fibre is of great importance. For many years there have been varying degrees of dissatisfaction with the "Weende" method of fibre determination which has been used in laboratories all over the world for well over a century. 'Crude fibre' is determined by boiling a sample of food (after fat extraction) successively with 1.25% H_2SO_4 and with 1.25% NaOH, each reagent being employed for exactly 30 minutes. 'Crude fibre' is the organic matter surviving this 'digestion.' The N.F.E. is estimated by subtracting the amount of crude protein, ether extract and crude fibre from the amount of organic matter in a sample and, thus, any errors in these three determinations will affect the accuracy of the derived value for N.F.E.

The fibrous components of plant materials are cellulose, hemicellulose and lignin. The Weende system of analysis fails to isolate all of these components in the 'crude fibre' fraction since the hemicellulose and part of the lignin dissolve in the digestion procedure used (Van Soest & Moore, 1965). Consequently, the crude fibre figure underestimates the total fibrous material (cell wall constituents) in plant material and so the figure for N.F.E. (calculated by difference) is likewise inaccurate in representing soluble carbohydrates. Despite the theoretical objections to the use of the Weende method of analysis, the parameters of a feed obtained are of considerable practical value. Crude fibre as measured in the previously-described way does show strong negative correlations with digestibility of the organic matter of feeds (See Nordfeldt, 1954) and digestibility of energy in diets (See Henry, 1971).

The influence of crude fibre on organic matter digestibility varies from feed to feed, depending on the special character of the crude fibre in individual feeds. Nordfeldt (1954) found, for example, that with barley diets a rise in crude fibre % of 3 units gave just over 1% depression in organic matter digestibility. For diets based on oats, a similar rise in crude fibre level gave a 3% depression in organic matter digestibility, and for wheat bran diets a rise in fibre % of 3 units was associated with a depression in digestibility of 10%.

The fibrous portion of a food, being fairly indigestible to a pig, influences the digestibility of the other constituents by exerting a 'protective' action, encasing these constituents in a digestion-proof shield, as it were. For efficient pig feeding, some form of physical treatment of cereal grains is essential to ensure the breaking down of the fibre encapsulating the more soluble constituents so that digestive secretions can penetrate more completely. At the present time grinding is the treatment which is most commonly used (often followed by pelleting of the complete diet). The degree of fineness of grinding appears less critical than was previously thought except, perhaps, for oats where fine grinding does show advantages (see Lawrence, 1976). In some experiments rolled grain has supported equivalent performance to ground grain (see Lawrence, 1972). In contrast to its beneficial effect on digestibility of cereal grains, fine grinding of dietary grains has been found to predispose towards the occurrence of gastric ulcers in pigs (e.g. Riemann, Maxwell, Kowalczyk, Benevenga, Grummer & Hoekstra, 1968).

Ideally, laboratory methods of fibre estimation should distinguish between the components of 'fibre' since lignin is indigestible while cellulose and hemicellulose are of low digestibility by non ruminants (Van Soest, 1967). Analytical methods for the partition of the fibrous components are already available (Van Soest & Moore, 1965) although a full partition does involve a larger amount of laboratory work than simple crude fibre or detergent fibre estimations. However, Henry (1971) found that 'acid detergent fibre' (A.D.F.) and 'neutral detergent fibre' (N.D.F.) measurements each showed stronger negative correlations with apparent digestibility of energy than did 'crude fibre'. When results from diets containing up to 15% lucerne meal were included, N.D.F. measurements predicted the energy content of faecal organic matter much better than either A.D.F. or crude fibre measurements. Since in the N.D.F. technique all the highly digestible "cell contents" of plant material are removed, the residue, containing hemicellulose, cellulose and lignin, would be expected, on theoretical grounds, to give a good index of indigestible fibre in rations for non ruminants.

THE EFFECTS OF MICROBIAL FERMENTATIONS AND INTER-CONVERSIONS ON CALCULATIONS OF DIGESTIBILITY OF INDIVIDUAL SUBSTANCES

The concept of digestibility assumes that the feed residues found in faeces are merely the result of inefficient digestion and nothing else. However, the position is more complex. Faeces contain undigested feed residues, unabsorbed digestive secretions, intestinal cells which are continually being sloughed off, dead micro-organisms and, last but not least important, the products of microbial fermentation. The unabsorbed residues of digestive secretions and cellular debris are predominantly protein in nature and their presence in the faeces has been discussed as "endogenous nitrogen" in Chapter 7. The bacterial deamination of amino acids is mentioned earlier in the same chapter. Fermentation of carbohydrates, described in Chapter 6, materially affects the interpretation of digestibility measurements. Because faeces rarely contain sugars, we cannot assume that sugars have always been completely digested and their hydrolysis products completely absorbed. In the very young pig, sugars will be fermented to some extent in the stomach before reaching the digestive part of the gut, since pH in the stomach soon after birth is relatively high and micro-organisms can multiply. Later, as acid secretion develops, such fermentations will be reduced in the stomach, but in the caecum and large intestine, throughout the life of the pig undigested sugars will be fermented rapidly. So a 'digestibility' figure for sugars is meaningless as an index of the percentage of a sugar which is broken down in the small intestine and, thus, is available for absorption. For fats, too, digestibility calculations for individual fatty acids could mislead if one were unaware of the activities of the gut microflora. Carlson & Bayley (1968) have shown that hydrogenation of linoleic and oleic acids occurs in the caecum and colon of young pigs and that calculations of digestibility for these fatty acids and for stearic acid (the end result of hydrogenation of the former acids) can give, and are likely to give, spurious results. High 'digestibility' for oleic acid and low 'digestibility' for stearic acid may just be the result of the microbial conversion of the former into the latter (see Chapter 8).

FACTORS INFLUENCING DIGESTIBILITY

Measurements of digestibility are only precisely valid for the conditions under which they are taken. For example, the digestibility of a particular protein will vary to some extent according to the level at which it is fed, will be influenced by other constituents of the diet, and may also depend on the age and other characteristics (e.g. breed) of the group of pigs used. Comparisons of digestibility are therefore usually more useful than the absolute values obtained, which might have been different if they had been measured with different pigs or under different conditions. Figures showing the digestibility of the components of different feedingstuffs can be found in books on practical pig feeding and differences in digestibility between feedstuffs will not be discussed here. Changes with advancing age in the ability of the pig to digest feed are of much more relevance to the theme of this book and evidence of such changes is outlined in the next section. Increases in ability to digest feed are not confined to the first few weeks of life when major changes in output of digestive enzymes are occurring. Improvements in digestibility appear to occur throughout the growing and fattening period and to continue beyond this stage in animals kept for breeding.

Changes in digestibility of foods with increasing age of pigs

The first two months of life

At birth, the pig's digestive system is immature. Changes in the output of digestive enzymes and secretions have been described in earlier chapters on protein, fat and carbohydrate digestion and these changes are reflected in rapid improvements in digestibility of certain nutrients, particularly during the first two months of life. In the first fortnight or so of life pigs require very highly digestible feed and would be receiving a diet of sow's milk supplemented with very small amounts of supplementary food, if reared naturally. Milk output by sows rises during the first 2 or 3 weeks of lactation and this increase in milk output tends to keep pace with the pig's increasing appetite over this period. After about the third week of lactation the sow's milk yield tends to remain steady, or, possibly, to fall. Thus, from 2 or 3 weeks of age, milk availability no longer keeps up with the increasing demands of the growing piglet. So, from the second or third week of age the piglet will show increasing interest in supplementary food, the intake of which will rise steadily. The gradual introduction of the piglet's immature digestive system to feeds which are much less digestible than sow's milk reduces the likelihood of digestive upsets arising from inability to utilise the supplement. Artificial diets to be used from a very early age need to be composed of milk products and easily digestible carbohydrates if rapid growth and efficient feed conversion is required. Diets consisting of cereals and a mixture of cheaper protein sources would tend to cause a depression or even a cessation of growth if used for early weaning at 7–10 days of age. The cost of milk replacer diets limits the length of time for which they are used, and as soon as the piglet is capable of digesting cheaper feed mixtures there is considerable economic pressure to replace milk-based diets with cereal and fish protein or similar, cheaper mixtures. Such a changeover will only depress growth if carried out too soon and might even precipitate outbreaks of gastroenteritis. The age at which the change may safely be made is not clearly defined and depends on many factors such as the precise composition of the cereal-based mix, the standard of hygiene and management in the rearing house, whether the pigs are on wire floors or, alternatively, in pens on the floor in contact with their own faeces.

The piglets within a batch which fail to adapt to a less digestible early-weaning diet will tend to be those whose digestive enzyme development is most retarded. If such pigs become ill because they cannot digest their food their presence in the group may act as a source of infection which may spread throughout the batch.

In experiments with pigs weaned at three weeks of age on to diets based on barley or maize, Aumaitre (1976) found that apparent digestibility of organic matter rose markedly between 3 and 5 weeks of age and continued to rise, more slowly, up to the 8th–9th weeks of life. Similar improvements with age were shown in apparent digestibility of nitrogen. The maize diets showed superior digestibility to the barley diets throughout the experimental period from 3–9 weeks of age. Frequency of occurrence of diarrhoea was high in the immediate post-weaning period and progressively diminished, pigs on the maize-based diets showing a lower incidence of diarrhoea in the 5–9 week period than those on the barley diets.

Age changes in digestibility of food by older pigs

The data of Thorbek (1975) give an illustration of changes occurring in the 'growing-fattening' period. In this work the digestibility of feed by pigs was measured in 8 successive periods as the animals grew from 20 to 90 kg live weight.

Changes in digestibility in the later life of the pig are shown by the results of experiments reviewed by Nordfeldt (1954), where data concerning a much wider weight range of pigs were brought together and summarized. Thus, Thorbek's data covers the conventional growing-fattening period while Nordfeldt's review includes animals of fairly mature weights. Thorbek's data is particularly interesting because her studies were carried out using six different ration combinations, namely – barley and skim milk powder, barley and protein mix*, maize and skim milk powder, maize and protein mix*, sorghum and skim milk powder and sorghum and protein mix* (The protein mix* consisted of 67% soya bean meal and 33% meat and bone meal). Where milk protein was used to supplement maize or barley in Thorbek's work, the digestibility of dietary energy showed little change over the 20–90 kg weight range studied. However, when the mixture of soya bean meal and meat and bone meal was the protein supplement, digestibility of dietary energy rose from an average of 83.7% when pigs averaged 23.1 kg to an average of 86.7% when they averaged 79.9 kg. This represents an increase of about half a percentage unit in the digestibility coefficient for every 10 kg increase in body weight over the range 20–80 kg. The digestibility of protein did not show consistent changes with age in Thorbek's work.

Nordfeldt (1954) reviewed 1560 digestibility experiments with 1698 pigs covering the period from 1900 to 1951. To demonstrate the influence of age (weight) of pigs on digestibility of nutrients he grouped the pigs into three categories: Group I (below 100 kg body weight); Group II (100–180 kg body weight) and Group III (above 180 kg body weight). Table 9.1 summarizes his results. For organic matter, N-free extract and crude fibre there was a consistent tendency for digestibility to improve with increasing weight of pigs. For ether extract digestibility the trend was in the reverse direction, i.e. for a decline in digestibility in the heavier pigs. The crude protein digestibility was greater in group II than in group I but for group III the average digestibility coefficient was lower than for group II, but not significantly so.

Table 9.1 *Summary of the effect of body weight on digestibility of nutrients by pigs (Adapted from Nordfeldt, 1954)*

Nutrients	Digestibility coefficients (%) Group			Differences in average digestibility coefficients of nutrients between weight groups		
	I below 100 kg bodyweight	II 100–180 kg bodyweight	III above 180 kg bodyweight	between groups I & II	between groups I & II	between groups II & III
Organic matter	81.6	83.7	87.5	2.1***	6.0***	3.8***
N-free extract	88.0	89.1	91.7	1.1***	3.7***	2.6***
Crude protein	77.9	79.1	77.1	1.2*	0.8	2.0
Ether extract	64.1	59.3	55.9	4.8***	8.2*	3.4
Crude fibre	23.2	25.7	41.3	2.5***	18.1***	15.6***

REFERENCES

Allen, M.M., Barber, R.S., Braude, R. & Mitchell, K.G. (1963) A metabolic crate and harness suitable for male growing pigs up to bacon weight. *J. Anim. Tech. Ass. XIV,* 103–110.

Aumaitre, A. (1976) Évaluation de divers traitements technologiques des céréales. IV. Influence du floconnage et de l'expansion de l'orge et du mais sur les performances du porcelet sevré a 21 jours: effets sur la digestibilité des éléments de la ration. *Ann. Zootech., 25,* 41–51.

Carlson, W.E. & Bayley, H.S. (1968) Utilization of fat by young pigs: fatty acid composition of ingesta in different regions of the digestive tract and apparent and corrected digestibilities of corn oil, lard and tallow. *Can. J. Anim. Sci., 48,* 315–322.

Crampton, E.W., Ashton, G.C. & Lloyd, L.E. (1954a) Improvement of bacon carcass quality by the introduction of fibrous feeds into the hog finishing ration. *J. Anim. Sci., 13.,* 327–331.

Crampton, E.W., Ashton, G.C. & Lloyd, L.E. (1954b) The effect of restricting feed intake of market hogs during the finishing period on the quality of the bacon carcass. *J. Anim. Sci., 13,* 321–326.

Frape, D.L., Wolf, K.L., Wilkinson, J. & Chubb, L.G. (1968) Modification to Shinfield metabolism crate. *J. Inst. Anim. Technicians, 19,* 61–64.

Henry, Y. (1971) Essai de prevision de la valeur en energie digestible des aliments pour le porc, a partir de leurs teneurs en constituents membranaires. *Journées de la Recherche Porcine en France, Paris.* Inst. Tech. du Porc, & I.N.R.A. p. 57–64.

Lawrence, T.J.L. (1972) Developments in cereal processing – growing pigs *In* Cereal processing and digestion. Tech. Publ. U.S. Feed Grains Council, London. 1972. pp. 77–105.

Lawrence, T.J.L. (1976) Some effects of processing on the nutritive value of feedstuffs for growing pigs. *Proc. Nutr. Soc., 35,* 237–243.

McCarthy, J.F., Aherne, F.X. & Okai, D.B. (1973) Use of HCl-insoluble ash as an index material for determining apparent digestibility with pigs. *Canad. J. Anim. Sci., 54,* 107–109.

Nordfeldt, S. (1954) Digestibility experiments with pigs. *Kungl. Lantbrukshögsk. Ann., 21,* 1–29.

Reimann, E.M., Maxwell, C.V., Kowalczyk, T., Benevenga, N.J., Grummer, R.H. &
Hoekstra, W.G. (1968) Effect of fineness of grind of corn on gastric lesions and contents of swine. *J. Anim. Sci., 27,* 992–999.

Thorbek, G. (1975) Studies on energy metabolism in growing pigs. II. Protein and fat gain in growing pigs fed different feed compounds. Efficiency of utilization of metabolizable energy for growth. *Beretn. Forsogslab., 424,* 193 pp. Kobenhavn.

Van Soest, P. J. & Moore, L.A. (1965) New chemical methods for analysis of forages for the purpose of predicting nutritive value. *Proc. 9th Int. Grassl. Congr.,* Sao Paulo, 783–789.

Van Soest, P.J. (1967) Development of a comprehensive system of feed analysis and its application to forages. *J. Anim. Sci., 26,* 119–128.

INDEX

UNIVERSITY OF BRISTOL VET. SCIENCE

78 20333